IITK Directions

Volume 4

IITK Directions is an institutional series of the Indian Institute of Technology Kanpur. IITK Directions aims at presenting original contributions and reviews of important and cutting-edge topics by the faculty members and other stakeholders of IIT Kanpur in a consolidated manner. It is a platform that reflects upon the ongoing research and development activities, major achievements, and evolving trends. Each volume in the series focuses on a particular area of science and technology with chapters written by the faculty members, students and research staff. Each chapter is written in a journalistic tone for the peer group – readability and accessibility being important parameters. The volumes contain a survey of the subject as a whole and the extent of contributions recorded by the Institute faculty and students who participate so extensively in research.
The volumes in this series will be useful to researchers, practitioners, students, industry leaders, government officials and policy makers. Above all, the series serves as platform to publish the most cutting edge research summaries and evaluate the applications and impact of the research work undertaken at IIT Kanpur.

More information about this series at http://www.springer.com/series/15345

Sandeep Kumar Shukla ·
Manindra Agrawal
Editors

Cyber Security in India

Education, Research and Training

Springer

Editors
Sandeep Kumar Shukla
Department of Computer Science and Engineering
Indian Institute of Technology Kanpur
Kanpur, Uttar Pradesh, India

Manindra Agrawal
Department of Computer Science and Engineering
Indian Institute of Technology Kanpur
Kanpur, Uttar Pradesh, India

ISSN 2509-6591 ISSN 2509-6605 (electronic)
IITK Directions
ISBN 978-981-15-1674-0 ISBN 978-981-15-1675-7 (eBook)
https://doi.org/10.1007/978-981-15-1675-7

This Springer imprint is published by the registered company Springer Nature Singapore Pte Ltd.
The registered company address is: 152 Beach Road, #21-01/04 Gateway East, Singapore 189721, Singapore

Series Editor's Preface

I am quite pleased about this special issue of Directions which has focused on research & development in the areas of cybersecurity and related technologies. At this point, I would also like to explain the genesis of the current issue.

An "Interdisciplinary Center for Cybersecurity and Cyber Defense of Critical Infrastructures (C3I Center)" was established in 2016 at IIT Kanpur with funding from the Science and Engineering Research Board (SERB), and other funding agencies. The objective of this interdisciplinary center is to build India's first industry-scale cybersecurity testbed and to carry out research on smart grid security, manufacturing, and other industrial control system security network. This issue, therefore, will highlight the work carried out at IIT Kanpur in this direction, and would also provide some perspectives from other Indian institutes on the research work on cybersecurity is being initiated. Overall, we intended to provide a collection of articles as a snapshot of the R&D efforts in the cybersecurity field which we believe will be of use to academic institutions as well as government agencies. We will be immensely satisfied if this endeavor can initiate new collaborations, both academic and industrial, in the near future.

I am thankful to the authors for their spontaneous participation to contribute these articles and make timely revisions based on editors' notes. I am also thankful to the editors and the publication team for raising the quality of the publication. It is worthwhile to note that this is the third issue of Directions which is published by Springer. Based on our experience with the earlier issues, we expect to achieve a wider circulation for this issue as well, and surely the quality of the two-stage editing process will be appreciated by the readers.

Prof. S. Ganesh
Dean, Research and Development
Indian Institute of Technology Kanpur
Kanpur, India

Contents

Editors and Contributors

About the Editors

Sandeep Kumar Shukla is currently the Poonam and Prabhu Goel Chair Professor and Head of the Computer Science and Engineering Department, Indian Institute of Technology, Kanpur, India. He is the Editor-in-Chief of ACM Transactions on Embedded Systems and an Associate Editor of ACM Transactions on Cyber-Physical Systems. He is an IEEE fellow, an ACM Distinguished Scientist, and served as an IEEE Computer Society Distinguished Visitor from 2008 to 2012, and as an ACM Distinguished Speaker from 2007 to 2014. He was previously an Associate Editor of IEEE Transactions on Computers, IEEE Transactions on Industrial Informatics, IEEE Design & Test, IEEE Embedded Systems Letters, and various other journals. He was a member of the faculty at the Virginia Polytechnic Institute, Arlington, Virginia, between 2002 and 2015, and has also been a visiting scholar at INRIA, France, and the University of Kaiserslautern, Germany. In 2014, he was named a fellow of the Institute of Electrical and Electronics Engineers (IEEE) for his contributions to applied probabilistic model checking for system design. He has authored several books on systems and has edited and co-authored numerous books with Springer.

Prof. Manindra Agrawal received his B.Tech. and Ph.D. in Computer Science and Engineering from the Indian Institute of Technology, Kanpur in 1986 and 1991, respectively. He was a fellow of the School of Mathematics, SPIC Science Foundation, Chennai, from 1993 to 1995, and a Humboldt fellow at the University of Ulm, Germany, from 1995 to 1996. He joined the faculty at IIT Kanpur as an Assistant Professor at the Department of Computer Science and Engineering in 1996. And was appointed as the N. Rama Rao Chair Professor in 2003. He is the recipient of several international awards, including the Fulkerson Prize 2006 and the Gödel Prize 2006, and has published and presented papers in respected journals. He was the Head of the Computer Science and Engineering Department, and Dean of Faculty affairs at IIT Kanpur. Currently, he is the Deputy Director and Officiating Director of IIT Kanpur. Prof. Agrawal has made significant contributions to the theory of efficient reactions between computational problems, which are part of the program studying the wellknown P vs NP question in mathematics/computer science. His joint paper with two of his former students resolves the centuries-old problem of a fast test of primality. In the language of complexity theory, they have proved that recognizing primes is in the 'class P' and this constitutes one of the most striking problems now known in this class.

Contributors

Shweta Agrawal Indian Institute of Technology Madras, Chennai, India

Ras Dwivedi Indian Institute of Technology Kanpur, Kanpur, India

Abhishek Gunda Indian Institute of Technology Kanpur, Kanpur, India

Mugdha Gupta Indian Institute of Technology Kanpur, Kanpur, India

Anand Handa Indian Institute of Technology Kanpur, Kanpur, India

Sai Krishna Kothapalli Indian Institute of Technology Guwahati, Guwahati, India

Amit Kumar Indian Institute of Technology Kanpur, Kanpur, India

Gaurav Kumar Indian Institute of Technology Kanpur, Kanpur, India

Nitesh Kumar Indian Institute of Technology Kanpur, Kanpur, India

Saurabh Kumar Indian Institute of Technology Kanpur, Kanpur, India

Nishit Majithia Indian Institute of Technology Kanpur, Kanpur, India

Devendra K. Meena Indian Institute of Technology Kanpur, Kanpur, India

Debdeep Mukhopadhyay Indian Institute of Technology Kharagpur, Kharagpur, India

Subhasis Mukhopadhyay Indian Institute of Technology Kanpur, Kanpur, India

Rohit Negi Indian Institute of Technology Kanpur, Kanpur, India

Biswabandan Panda Indian Institute of Technology Kanpur, Kanpur, India

Shubham Sahai Srivastava Indian Institute of Technology Kanpur, Kanpur, India

Rohit Sehgal Indian Institute of Technology Kanpur, Kanpur, India

Sanjay Sharma Indian Institute of Technology Kanpur, Kanpur, India

Sandeep Kumar Shukla Indian Institute of Technology Kanpur, Kanpur, India

Ajay Singh Indian Institute of Technology Kanpur, Kanpur, India

Shubham Singh Indian Institute of Technology Kanpur, Kanpur, India

Nilesh Vasita Indian Institute of Technology Kanpur, Kanpur, India

Building India's First Cyber-Security Test-Bed for CI

1

Rohit Negi and Sandeep Kumar Shukla

Abstract

Critical infrastructures such as power grid, water treatment plants, chemical plants, oil and gas refineries, transportation systems, etc., are examples of large-scale cyber-physical systems. In the recent years, there has been multitude of cyberattacks on such systems such as STUXNET attack in Iran, BlackEnergy malware and ransomware attacks on power distribution systems in Ukraine, and malware attacks on German Steel plant to mention a few. Critical infrastructure security is a major area of concern for all governments. In the United States, several laboratories such as Sandia National Lab, Idaho National Lab, and National Institutes of Standards have built critical infrastructure cyber-security experimentation test-beds. These test-beds are built with industry-scale equipment from various sectors of critical infrastructures to allow researchers to detect vulnerabilities in multitude of products that are used by utility companies in their systems and to research efficacy of their remediation techniques. These test-beds also allow utilities to bring in hardware and software products for hardware/software-in-the-loop testing for vulnerability detection or cyber-security product validation. In India, there has been no such test-bed for such research and validation activities. In the interdisciplinary center for cyber-security and cyber defense of critical infrastructures (C3i) at the Indian Institute of Technology Kanpur, we are building such a facility—a first in India. We already have built a lab-scale test-bed for power distribution automation and discovered a plethora of vulnerabilities in the architecture, protocols, and widely used products in our test-bed. In this article, we first describe the lab-scale test-bed to provide an insight into the utility of such a test-bed, and then we describe our ongoing development of industry-scale test-beds.

Keywords

Industrial critical systems • Critical infrastructure • Cyber-Security of Critical Infrastructure • Cyber-Security of Cyber Physical Systems

1.1 Introduction

As per today's scenario, power grid, water and sewage system, nuclear plants, industrial control of manufacturing plants, railway signaling and track switching systems, air traffic control and rockets/missile control—all are done through digital sensing and software-based control. Network-based data communication from sensors to control centers, substations, and load dispatch centers are examples of dependence of critical infrastructure on pervasive communication networks. Since the last few years, the usage of commercial of the shelf (COTS) hardware and software was introduced in entanglement with critical infrastructures. During the implementation and integration of COTS hardware/software, cyber-security has not been taken seriously and is not considered effective.

In the last decades of the previous century, most critical infrastructures started deploying SCADA systems. As embedded sensors, actuators, and centralized monitoring and supervision of system state for cyber-physical systems started becoming commonplace, SCADA-based automation became a necessity. More recently, due to IP convergence, SCADA systems started exchange of information with business network. An upcoming trend is to deploy web and agent-based SCADA systems [1] (Fig. 1.1).

In the recent few years, a number of incidents of cyber-attacks on critical infrastructure have been observed. A list

This work is funded by a research fund from the Science and Technology Research Board (SERB), Government of India.

R. Negi · S. K. Shukla (✉)
Indian Institute of Technology Kanpur, Kanpur, India
e-mail: sandeeps@cse.iitk.ac.in

R. Negi
e-mail: rohit@cse.iitk.ac.in

S. K. Shukla and M. Agrawal (eds.), *Cyber Security in India*, IITK Directions 4,
https://doi.org/10.1007/978-981-15-1675-7_1

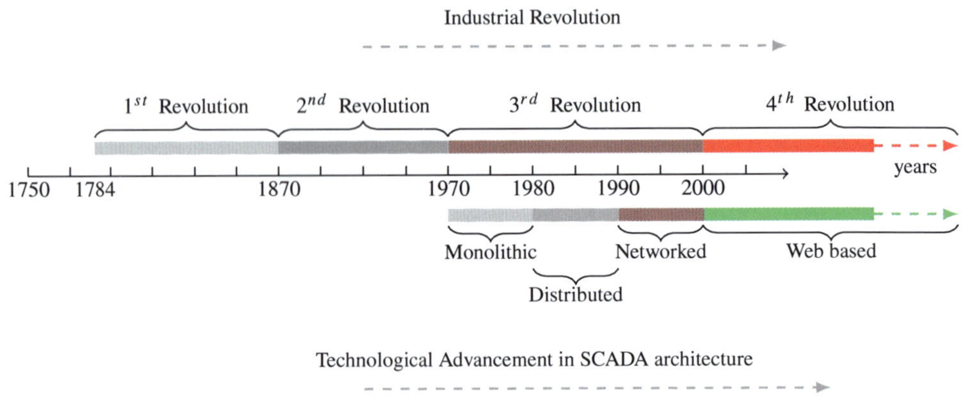

Fig. 1.1 Industrial revolution with SCADA advancement

of reported/identified cyberattacks is listed in Table 1.1. As critical infrastructures are playing an important role in economy of the nation, cyber-security and cyber defense of critical infrastructures have become an urgent priority. Nation-states, terrorists, and organized criminals can launch cyber-attacks on such systems to debilitate a nation's infrastructure, cause large-scale blackouts, manufacturing loss, train and air accidents, and nuclear accidents. So, cyber-security is no longer just an information security problem. It has become a national security problem due to increased use of digital control and communication in the functioning of critical infrastructures.

Usage of SCADA systems and operational technologies of critical infrastructure integrated with the business systems (IP convergence) makes the CI more connected and more exposed [2], and gives rise to new security risks [3]. CI becomes a lucrative target which attracts malicious scripts/intentional attacks.

In [4], a comprehensive analysis of the 242 cyber-security incidents on critical infrastructure reported between 1982 and 2014 leading to 673 deaths and 419 injuries has been presented. There are several cases reported after 2014 and some major ones are summarized in Table 1.1.

Almost all governments are in the process of building critical infrastructure security plans. For example, *National Critical Information Infrastructure Protection Centre of India (NCIIPC)* [18] was established in India in 2013 by an act of the parliament. In the United States, the *US Homeland Security of Critical Infrastructure Directive 7*, in Europe, European Programme for Critical Infrastructure Protection (EPCIP) [19], Centre for the Protection of National Infrastructure (CPNI) [20], etc., were established.

Since critical infrastructures tend to minimize downtime, and in situ penetration testing, vulnerability detection, or even patch-test-patch cycles cannot be done, one needs to resort to virtual or physical test-beds which either simulate or emulate the real infrastructure with all its instrumentation, networking, and control. Not only such test-beds are of importance for operational testing, but also a state-of-the-art test-bed helps researchers to study the threat vectors and develop mitigation techniques to avoid any undesired scenario by cyberattacks.

1.1.1 Contribution of This Article

In this article, we first describe our current test-bed which is a power distribution automation test-bed and illustrate its architecture. We also depict a few use cases which led us to discover some major vulnerabilities in the SCADA infrastructure provided by an internationally reputed vendor of SCADA software, relay, PLC, and protocol switches. Some of the vulnerabilities are due to the use of industrial protocols such as Modbus, and some are vendor-specific. This illustrates why one needs such test-beds, and simulation studies such as VSCADA [21] are not sufficient. Interestingly, these equipment that we have penetrated are in many real industrial and critical installations throughout India. Then we describe the industry-scale test-bed which encompasses power generation, transmission, and distribution subsystems, industrial automation test-beds, water treatment plant, and home automation. The test-bed components are at various stages of procurement and will be operational by 2020. So we are unable to provide any results in this article, but the infrastructure, architecture of the test-bed, and the span and scope of the test-bed provide a very detailed picture of what is being built at our center at IIT Kanpur.

This article is organized as follows. Sections 1.2 and 1.2.1 describe the C3i center and its objectives. Section 1.3 explains the existing experimental setup in our center. Section 1.4 elaborates the future planning of smart infrastructure.

1.2 C3i Center, IIT Kanpur

C3i center aims at spawning initiatives to develop technology and deploy technological safeguards to protect critical infrastructures. The goal of the center is to create India's first research center whose mission is research, education, training in the field of critical infrastructure protection and vulnerability studies. Science and Engineering Research Board (SERB) under the department of science and technology (DST) of the government of India, funded the "Interdisciplinary Centre for Cyber-Security and Cyber Defence of Critical Infrastructures" at IIT Kanpur, in March 2017.

Table 1.1 Cyberattacks on ICS

2018	Cryptocurrency Attack on Waste water site, USA [5]
2018	Triton Engineering Workstation, Undisclosed Worm [6]
2017	WannaCry Attacks the Nissan Car manufacturer, UK [6]
2017	NotPetya Multiple targets, Ukraine and Germany Ransomware [7]
2017	WannaCry infection suspected in Dacia production in Romania [8]
2017	CrashOverride The Malware That Took Down a Power Grid [9]
2016	SFG malware discovered in European energy company [10]
2016	U.S. Indicts 7 Iranians in Cyberattacks on Banks and a Dam [11]
2016	Conficker KGG Nuclear Power Facility, Germany Worm [12]
2016	Cyber-Attack on U.S. Water Facility [13]
2016	Encrypted firmware in PLCs by a ransomware affected Ukraine power grid [14]
2015	BlackEnergy malware to compromised the network controlling the power distribution shutdown caused blackout in Ukraine [15]
2015	Cyber Attack on German Steel Mill [16]
2015	Korean Energy and Transportation Targets Attacked by OnionDog [17]

The center is building India's first cyber-security test-bed for critical infrastructure, similar to what is available at Idaho national labs [22], Sandia National Labs [23], etc., in the United States. The researchers from our center in IIT Kanpur focus on discovering cyber threats to various critical infrastructure, developing solutions, and alerting the National Cyber-Security Coordinator's office, as well as other government agencies on the discovered vulnerabilities in our critical infrastructure sector.

The center is also playing an important role in enhancing cyber-security awareness by running online course security mooc powered by Mookit [24], organizing events of cyber-security awareness [25], and providing training at international level [26]. The center also trains students on this topic and technologies who would be cyber-security professionals in leading India's critical infrastructure utilities, and government agencies in the future. We are also planning to provide hands-on training to executives of utilities so that they can be aware of cyber-security threats, and how to prevent such attacks. The center is also engaged with international partners from Israel and USA in developing research and technology exchange, student training, as well as hosting conferences, workshops, and cyber-security competitions to create awareness, student excitement in choosing cyber-security profession, and to develop a world-class academic research culture in the field of cyber-security.

Fig. 1.2 Power distribution system with PLC/RTU-based SCADA system

In early 2017, we commissioned a flexible SCADA test-bed. The test-bed is one of the first of its kind in India. This test-bed is highly equipped with industrial hardware and software to provide a facility for research scientists and engineers investigating cyber-security of critical infrastructures. This test-bed replicates of power distribution automation for both three-phase and single-phase power supply. This test-bed is equipped with real field devices such as energy meters, protection relays, modular control system, IRIG-B-based GPS time sync unit, IEC-61850 compliant Ethernet switches, unified threat management hardware, industrial-grade HMI, SCADA system, etc. Communication between protection and control system up to the SCADA host can be configured via several protocols (Fig. 1.2).

During experiments, it has been observed that industrial hardware is highly vulnerable and can be exploited with reasonable ease. Attacks like DOS, MITM, privilege escalation, and control hijack have already been executed on the same.

1.2.1 Objective

It is essential that the research work be undertaken in realistic and accurate conditions, so that the subsequent benefits found can be reproduced in real-life scenarios. C3i center will provide these conditions within a controllable and realistic environment, using the latest technology that can be found in smart grid, smart home, smart transportation system, automated manufacturing and process lines. Current focus is on cyber-security related to system software, networks, mobile platform, supply chains, etc. Finding more ways to exploit, to expose the attack surfaces and discover newer threat models, as well as to find the mitigation techniques.

In future, C3i center will provide a valuable demonstration facility in a wide variety of disciplines for learning about the building blocks of smart infrastructures. It will address the challenging task of research and development of security modules for critical infrastructures. It is required to understand the complexity of critical infrastructure and to provide a state of the art test-bed facility for research scientists and engineers to investigate on various topics like:

- **Embedded and cyber-physical system,**
- **Protocol reverse engineering and lightweight encryption,**
- **Information security module,**
- **Intrusion detection and prevention modules,**
- **Honeypot,**
- **Industrial automation,**
- **Industrial Internet of things,**
- **Blockchain,**
- **Formal verification, and**
- **Memory systems security.**

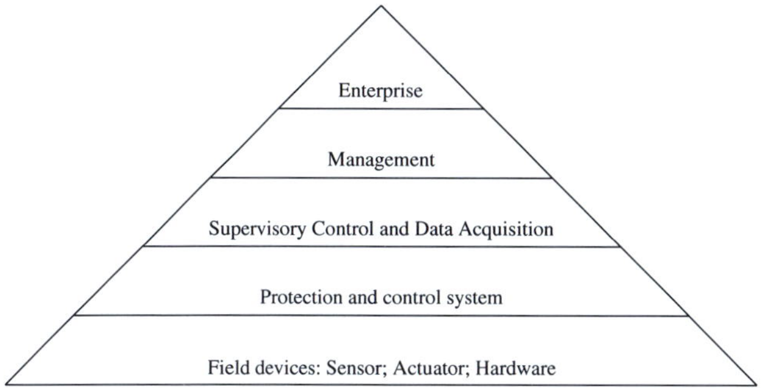

Fig. 1.3 Industrial Automation Pyramid

1.3 Existing Setup: Experimental Setup

The industrial automation is constantly evolving with the integration of operation technologies and information technology.

This integration of technologies is represented well by "Industrial Automation Pyramid" that includes the five technological levels as depicted in Fig. 1.3. The technologies interrelate both within each level and between the different levels by using industrial communications [27].

One of the largest installed base of industrial automation is found in the power generation, transmission, and distribution systems. C3i center has power distribution test-bed which is designed and implemented as a flexible, multilevel, hierarchical structure and includes four levels as mentioned in Table 1.2 to set up a realistic industrial automation environment. In the distribution subsystem of a power grid, for example, one would have automation units, protective relays, distribution feeders, and distribution substations. For the field emulation, we created a replica of a power distribution and feeder automation system. With this field itself, we use several standard field data communication protocols (Fig. 1.4).

1.3.1 Level-0: Field Devices

In our field panel, we have one incoming and five outgoing feeders as illustrated in the single line diagram as shown in Fig. 1.5. Emulation of both single-phase load and three-phase load are possible on the field panel. Conditions, such as overvoltage, undervoltage, time defined overcurrent, instantaneous overcurrent, connection failure, load failure, etc., can be emulated. Brief description of these outgoing connections is as follows:

- Outgoing-1 is to connect with three-phase load. Three single-phase variacs (variable transformers), one for each phase, are used to emulate phase-specific faults.
- Outgoing-2 is to connect with three-phase load. One three-phase variac is used to emulate three-phase faults.
- Outgoing-3, 4, 5 are to connect with single-phase loads. Single-phase variacs are used to emulate faults single-phase faults.

Each of the above three items is referred as a *section* of the field panel. Thus, the field panel has three sections which can be operated either independently or operated together.

Incoming connection of the field panel is connected with a 11KV/415V substation of the institute and every outgoing of the panel is feeding load for emulation. Each phase of every outgoing line consists of circuit breakers (XCBR), instrument transformers, i.e., potential transformer (PT) and current transformer (CT), contactors with overload relay and auxiliary contacts. These instruments are further integrated with metering units, PLC, and protective relays.

Expanding on Figs. 1.5, 1.6 illustrates an experimental setup where an outgoing connection is feeding an inductive load (i.e., a motor). As per the relay logic, there are start/stop push buttons for controlling the motor and overload trip relay for protection of the same.

This power distribution panel is integrated with industrial automation. The panel is installed with selector switches, selection modes of which decides the mode of operation between local, remote, and maintenance modes. The different modes of the system are affected by an interlock circuit as illustrated in Fig. 1.7. If the selector switch is positioned in the maintenance mode, then there is no option to run the system. In the local mode, interlocks are manually actuated using the start and stop buttons on the field panel. In the remote mode, control of the field is transferred to the control system and all interlocks are activated either semiautomatically or fully automatically. In semiautomatic mode, an operator intervention is required for actuating field devices and modifying set-points.

The remote mode has two sub-modes, namely, auto and manual modes. In the remote-auto mode, the controller responds to all field events where an action is required as per the control logic. In the remote-manual mode, the controller will be reading the field values but waits for an operator to respond to any actionable event.

1.3.2 Level-1: Protection and Control System

This level consists of a controller which drives the contactors using the information fed by field instruments. At this level, PLC and RTU modules and protection relays are implemented. Push buttons, selector switches, contactors, and metering units all are integrated with the protection and the

Table 1.2 Level in industrial automation

Level	Category
Level-0	Field devices such as potential transformers, current transformers, circuit breakers, energy meters, contactors, relays, etc., mounted on electromechanical infrastructure
Level-1	Protection and control system of the same facility
Level-1.5	An intermediate level between level-1 and level-2 that covers communication protocols such as IEC-61850 [28], IEC-60870-5-104 [29], Modbus TCP, Modbus RTU [30], NTP [31], IRIG-B [31], etc.
Level-2	Both centralized and localized real-time monitoring, data logging, and supervisory control of the process
Level-3	Historical report generation and publication over the network for analysis and audit

control system. The field instruments share real-time information about their status, voltage, and current measurements via hardwired signals or over Modbus RTU using RS485-based serial communication.

1.3.2.1 PLC/RTU-Based Control System

A modular approach facilitates emulation of both process automation and energy automation scenario. Single rack with modules of 24V DC power supply, PLC module, RTU module, 32-channel digital input, 16-channel digital output, and 8-channel analog input is implemented. However, it has space for future expansion. It can support up to 1024 digital input/output channels and 256 analog input/output channels. This control system can be programmed in several programming paradigms: ladder logic (LD), functional block diagram (FBD), structure text (ST), sequential function chart (SFC), and instruction list (IL) by using proprietary IDE (Unity Pro).

In general, controller interprets input signals and carries out the control actions as per the control philosophy. To emulate power distribution and feeder automation, RTU module is configured to communicate with SCADA host PC using IEC-60870-5-104. To emulate a process automation, PLC module is configured to communicate process information to SCADA host PC using OPC (Table 1.3).

1.3.2.2 Feeder and Motor Protection Relay

A protective relay senses power-system disturbances and automatically performs desired actions on the power system to protect the equipment. It provides comprehensive protection capabilities, including time overcurrent, definite-time overcurrent, over/undervoltage, circuit auto-reclosing, over/underfrequency, etc. Using an IDE (VAMPSET: a relay management software), this protective relay is configured and integrated either with the control system or with SCADA host computer via IEC 61850, IEC 60870-5-101, Modbus TCP, DNP 3.0 and Ethernet IP. Currently, this is configured on IEC 61850 protocol to read or write static data or to receive events sent spontaneously from the relay to SCADA host computer.

1.3.2.3 Time Stamping and Sequencing

A geographically distributed system like a power system requires all the automation units synchronized to the accuracy of nanoseconds. All the geographically distributed data will be time stamped so that faults during system outages can be located easily. In our test-bed, use of IRIG-B and NTP for clock synchronization allows reducing measurement uncertainty and producing meaningful results.

1.3.3 Level-1.5: Industrial Communication

A major feature of this test-bed is the detailed implementation of industrial protocols in between level-1 and level-2. Table 1.4 is listing the communication protocols used in the test-bed with nodes A and B corresponding to the devices used at the different levels of the system. This is a critical part of the test-bed, as the network is often the first target to be exploited, and can be exploited remotely. Many hostile conditions can be avoided through implementing a secure network. The vulnerability of protocols can be demonstrated through the emulation of real-world protocols on industrial hardware and software rather than the simulation of the protocols on virtual SCADA we had built earlier.

1.3.4 Level-2: Visualization and Control

This level drives the automatic control system by setting target or goals of the controller. Supervisory control looks after the equipment, which may consist of several local control loops. It serves to translate complex process variable values into usable and actionable information.

1.3.4.1 Supervisory Control and Data Acquisition Software

SCADA software acts as a *Centralized User Interface*. A PC-based system is used due to high demands on the quantity and type of information that must be processed and documented. It

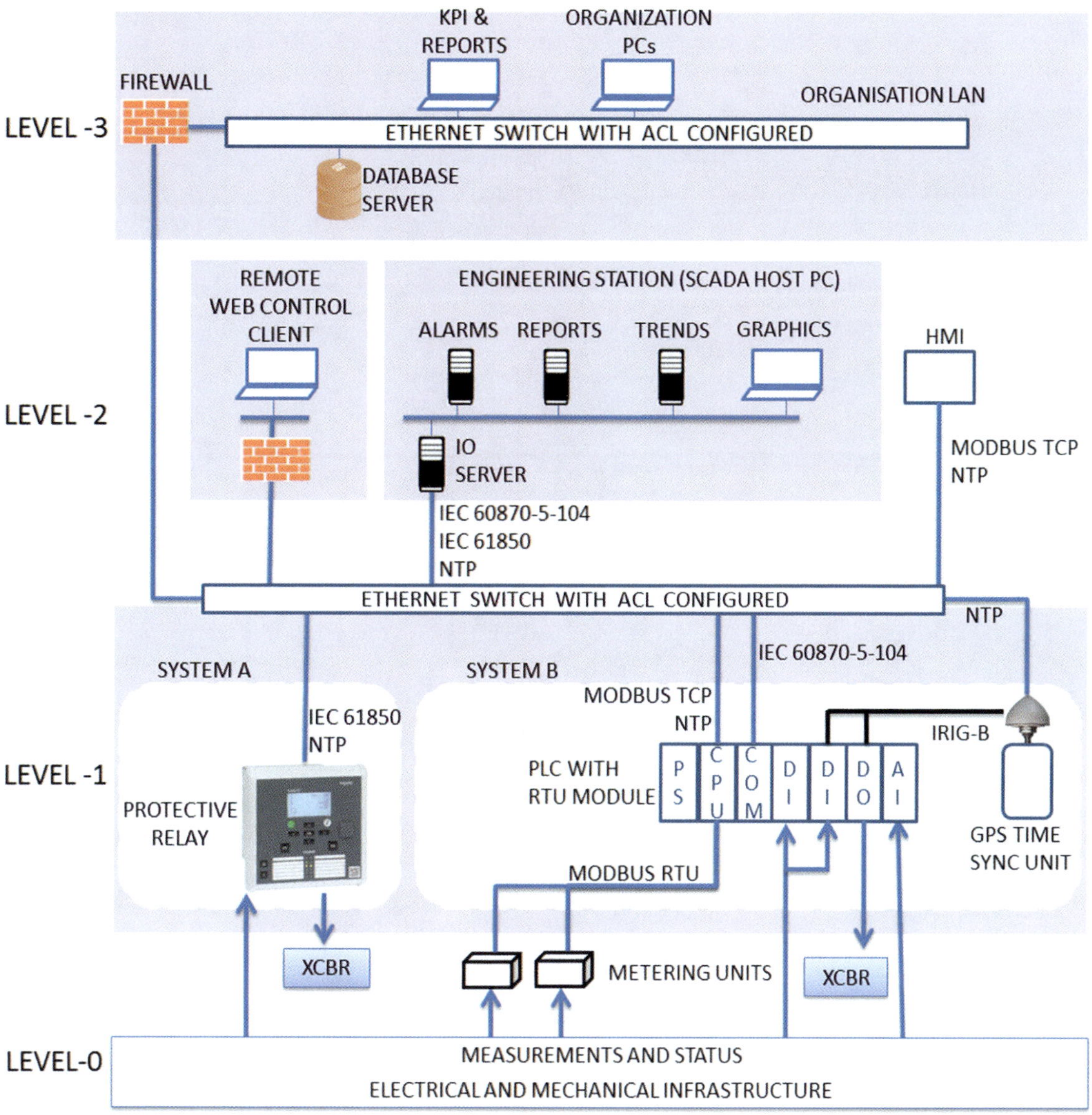

Fig. 1.4 Control architecture of a *section* in the field panel

offers options for sufficient storage space, processing power, and data connectivity. Any application that gets operating data about a process in order to control and optimize the process is a SCADA application. Applications differ based on the industry verticals such as oil and gas, water/wastewater treatment, manufacturing, etc. We are using Schneider made SCADA software. Major sections of SCADA host computer are shown as engineering station in Fig. 1.4. Here the IO server (data acquisition server) is communicating with RTU module over IEC 60870-5-104 and with protective relay (numerical relay) over IEC 61850. Further, it serves field data to remote web clients and runtime applications. To emulate web-based control clients, web-based SCADA system is incorporated with the same test-bed.

Life cycle of commercial-grade PCs is short and they are not reliable for field, i.e., harsh environment conditions. Availability of the system is a primary concern. Therefore, robust hardware for tough and harsh environment is required. Panel mounted industrial-grade PCs are recommended. However, industrial-grade PCs are expensive and high cost of a SCADA software would make the project over budget. So hardware-dependent HMI software is widely used. They are cost-effective and same has been implemented in test-bed for decentralized user interface. It provides the interface between the process and the operators—in essence an operator's dashboard. It acts as a primary tool by which operators and line supervisors coordinate and control the industrial processes in the plant floor.

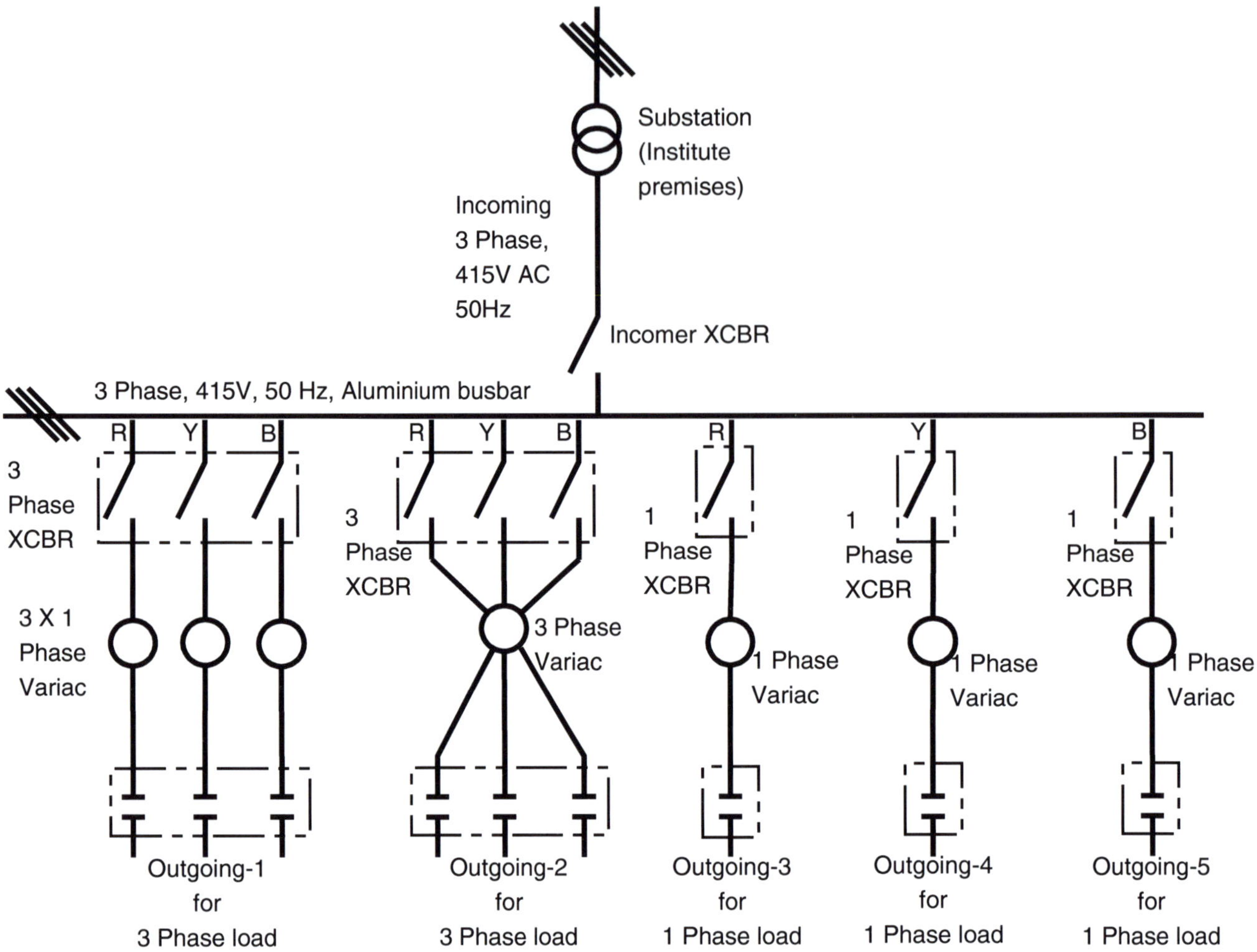

Fig. 1.5 Single line diagram of power distribution system having one incomer and five outgoings

Table 1.3 Example IO list

S. No.	Category	Description
1	Digital input	XCBR close feedback
2	Digital input	XCBR trip feedback
3	Digital input	MAINTENANCE MODE feedback
4	Digital input	REMOTE MODE feedback
5	Digital input	LOCAL MODE feedback
6	Digital input	CONTACTOR open feedback
7	Digital input	CONTACTOR close feedback
8	Digital input	OVERLOAD trip feedback
9	Digital output	CONTACTOR close command
10	Digital output	CONTACTOR open command
11	Analog input	Voltage level
12	Analog input	Running load
13	Soft IO	Modbus RTU communication

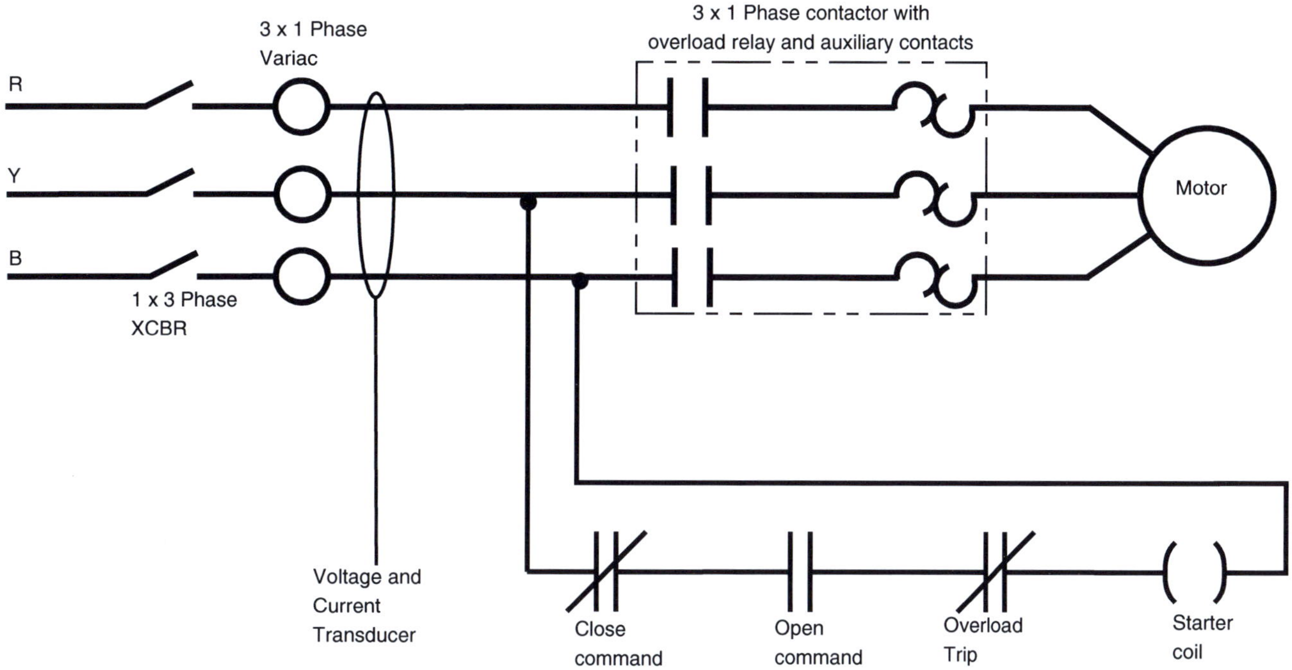

Fig. 1.6 Control architecture of a *section* in the field panel

Table 1.4 List of protocols available on test-bed

S. No	Communication protocol	Between nodes	
		Node A	Node B
1	IEC 61850	Protection relay	SCADA host computer
2	DNP 3.0	Protection relay	SCADA host computer
3	Modbus TCP	Protection relay	SCADA host computer
4	Ethernet IP	Protection relay	SCADA host computer
5	IEC 60870-5-101	Protection relay	SCADA host computer
6	IEC 60870-5-104	RTU module	SCADA host computer
7	Modbus TCP	PLC module	HMI
8	Modbus RTU	PLC module	Metering unit
9	OPC	PLC module	SCADA host computer
10	NTP	GPS TIME SYNC	Network devices
11	IRIG-B	GPS TIME SYNC	Digital input module

1.3.5 Level-3: Management

1.3.5.1 Data Logging for Reporting and Analysis

Schneider made SCADA software supports integration with databases. In our laboratory environment, we have configured it with MS SQL Server. This facilitates storing logs and analytics data which gets processed to publish historical reports and trends to be used at Level 3 as shown in Fig. 1.4. Note that even historical data ex-filtration could be targeted by cyberattacks—as one could estimate many system parameters from such data, and use subsequently in constructing calculated attacks on the physical process and its stability.

Recently, lot of researchers are emphasizing on anomaly-based intrusion detection for cyber-physical systems that will enable us to detect novel or zero-day attacks unlike signature-based IDS. Lot of machine learning and deep learning techniques are used to develop anomaly-based IDS. But there are very small number of datasets available for the researchers to

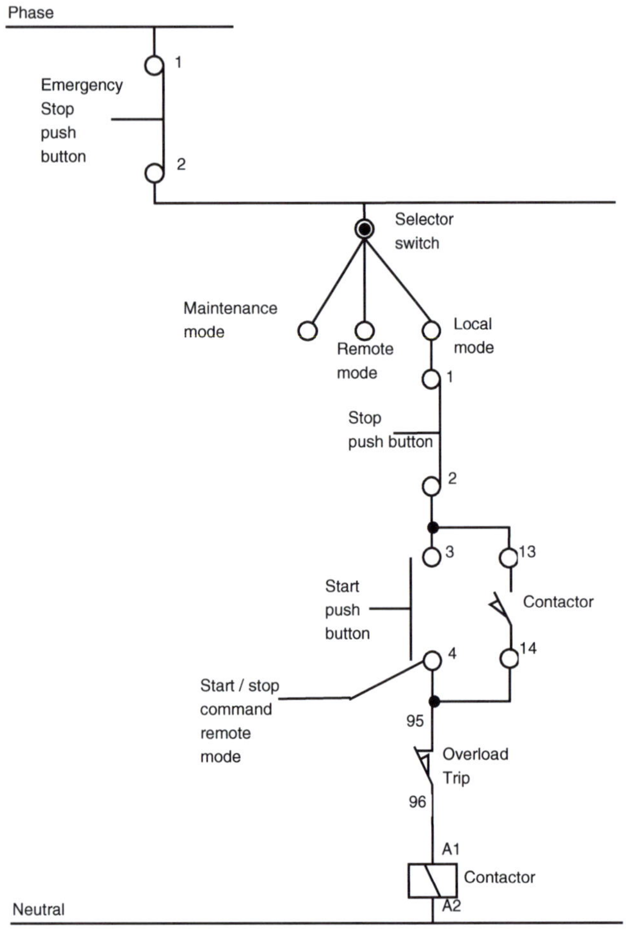

Fig. 1.7 Interlock control mechanism with PLC. Initially, a permissive check is carried out for the presence of any request for emergency shutdown and an overload trip. If passed, the mode of operation depends on the position of the selector switch. When the overload condition or emergency shutdown occurs, the interlock system is triggered for protection

work on. This logged time series data can be useful for the researchers to develop learning algorithms that will model the behavior of the cyber-physical system.

Availability of the ICS is the key requirement for any entity. So shutdown for routine patch management is not possible. Real-time operating systems require specific memory, so hardware comes with limited space, any signature/anomaly-based detection is not possible at controller level. Industrial hardware/software are very expensive, modification with the same is again a big deal. Budget, space, resource constraints are always there in the plant environment.

Plant floor people require secure wrapper for ICS with zero shutdown, by which they can increase the productivity without any fear. They require automated notification along with threat intelligence system of the entity and a repository of all well-defined threats for analysis and enhancement of security (Table 1.5).

1.4 Future Planning of Smart Infrastructure

A smart infrastructure is a complex integrated infrastructure with real-time monitoring, coordination, and synergistic functioning of different participants like power, oil, gas, transportation, water, drainage, and sewage. It provides sustainable economic development and high quality of life by excelling in multiple key areas like economy, environment, energy efficiency, mobility, governance, people, and living conditions [32].

As depicted in Fig. 1.8, similar complex infrastructure will be replicated with different participants like smart power grid, smart water grid, smart industry, smart home, smart asset, smart transportation, smart farms, etc., to do research, education, training to protect critical infrastructure.

1.4.1 Smart Power Grid

A smart grid is a digitized power grid that allows a two-way communication between the end users, power generation, and intermediary stages such as transmission and distribution [33]. C3i center has a smart grid test-bed completely equipped with control, protection, measurement, and monitoring system. As depicted in Fig. 1.9, smart grid test-bed has three major components; they are smart generation, smart transmission, and smart distribution.

1.4.1.1 Generation

There are various sources of power generation among which a few are solar, diesel, wind, etc. C3i center has solar power plant, diesel power plant, and a power generation setup of three-phase motor as a prime mover with three-phase synchronous generator equipped with advance generation protection and control system which facilitates users by giving hands-on experience of power generation and its control system. The power generation units available with the C3i center are replicating distributed energy resources generating three-phase 415 V 50 Hz power and feeding the grid.

1.4.1.2 Transmission

The generated power is forwarded to power grid via transmission lines. The C3i center has step-up transformation setup and 25 KM long artificial transmission lines [PI-model using 6 Zebra conductor lines] equipped with distance protection. The users can create symmetric and non-symmetric faults using a fault injection module.

1.4.1.3 Distribution

The power distribution system carries power from transmission lines up to end users. C3i lab has power distribution

Table 1.5 IP addresses of devices

S. No	Devices	IP	Remark
1	SCADA host	192.168.100.56	
2	RTU	192.168.100.10	
3	HMI	192.168.100.115	
4	PLC	192.168.100.40	
5	Ethernet switch	172.27.18.222	Management IP of switch
6	Attackers host	192.168.100.15	Via DHCP
		172.27.18.220	Manual setting to access Ethernet switch

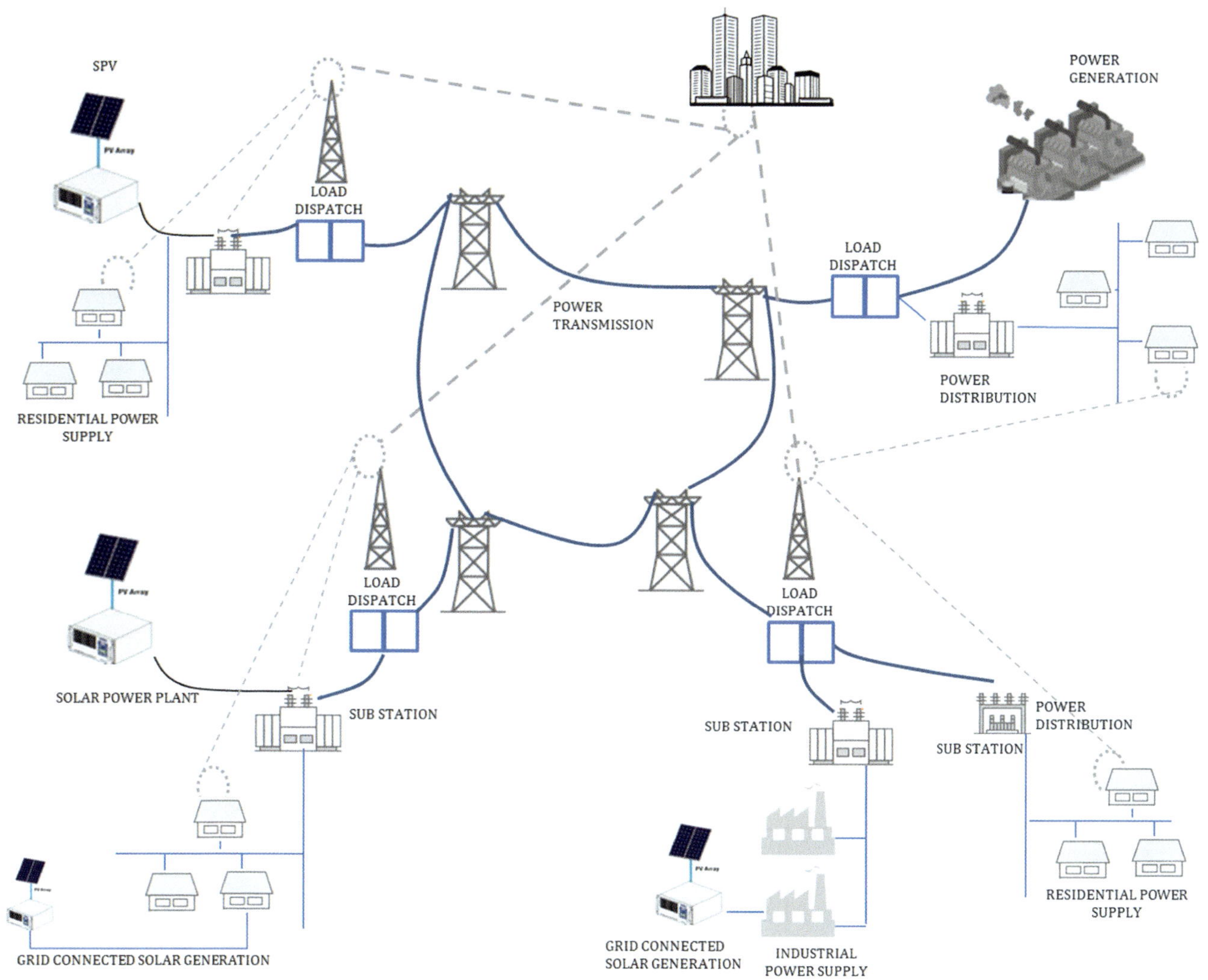

Fig. 1.8 Proposed architecture of smart infrastructure

system highly equipped with protection and control system, which includes transformation setups to step down the incoming power supply for power and motor control centers to feed the end users.

1.4.1.4 End User

C3i lab has diverse end users which will act as resistive, inductive, and capacitive load. C3i center has an advance load bank setup in which users can test smart grid behavior with balance

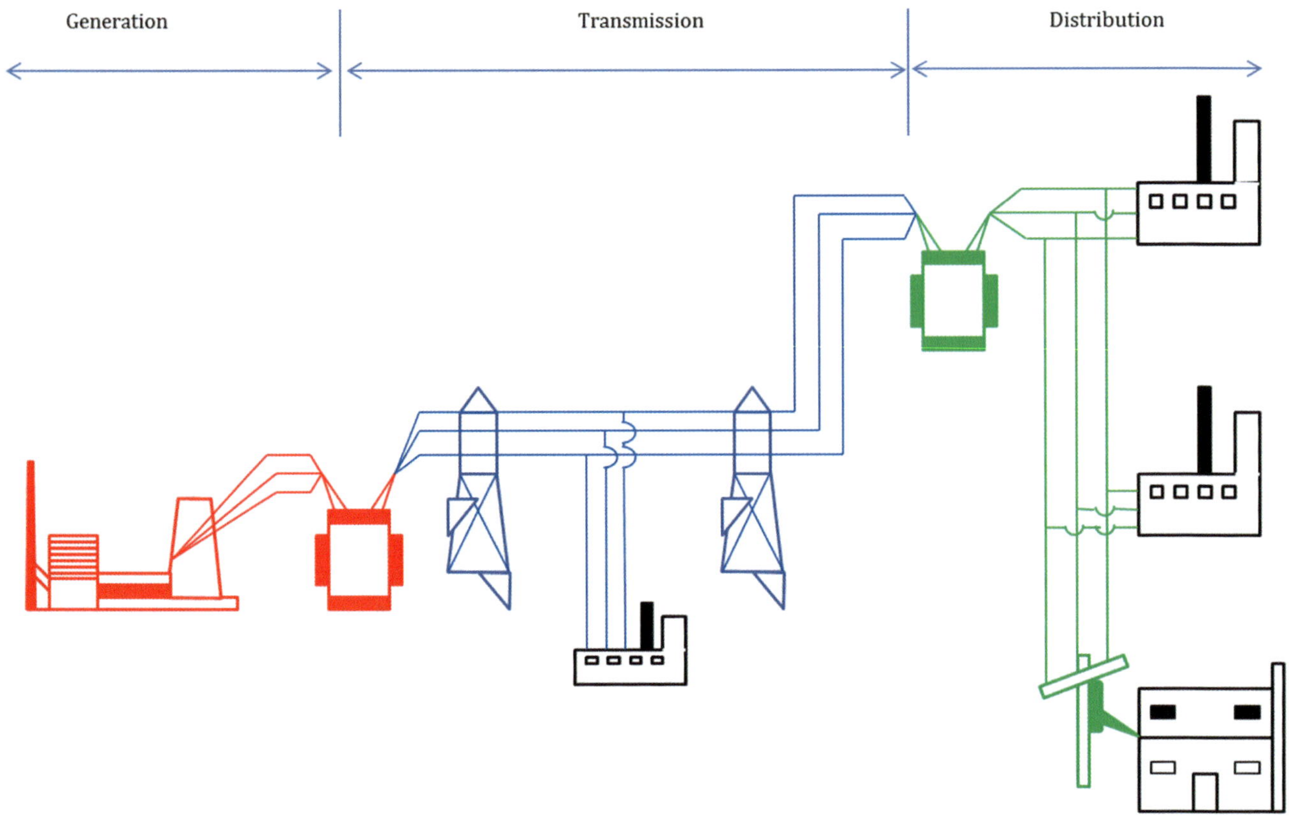

Fig. 1.9 The proposed architecture of smart grid test-bed

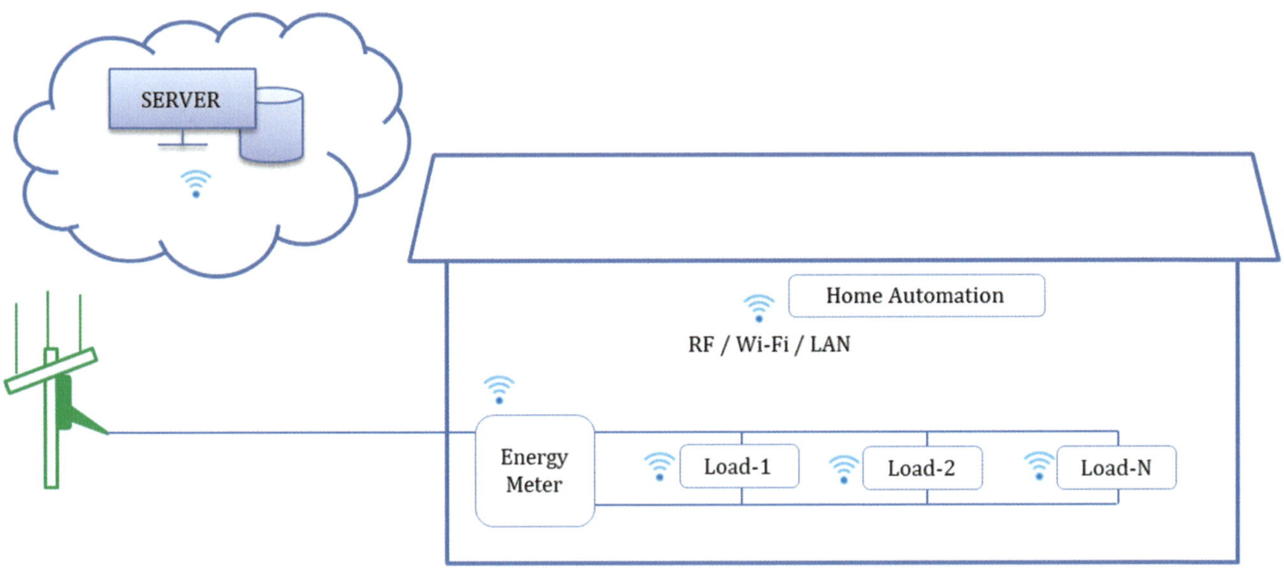

Fig. 1.10 Proposed architecture of home automation

and unbalance load. Along with advance load bank, C3i center has real electric load connected to smart power grid like smart home, smart water grid, smart industry, and smart farms.

Smart Home

A smart home is a fully automated home in which home automation system facilitates its owner by controlling climate, lights, and other home appliances [34]. As depicted in Fig. 1.10, C3i center has HAS (home automation system) which controls lighting, heating, cooling, etc. HAS optimizes the electricity bill by scheduling running hour of equipment. Also, it calculates total demand with respect to schedule and forwarded future demand to power grid.

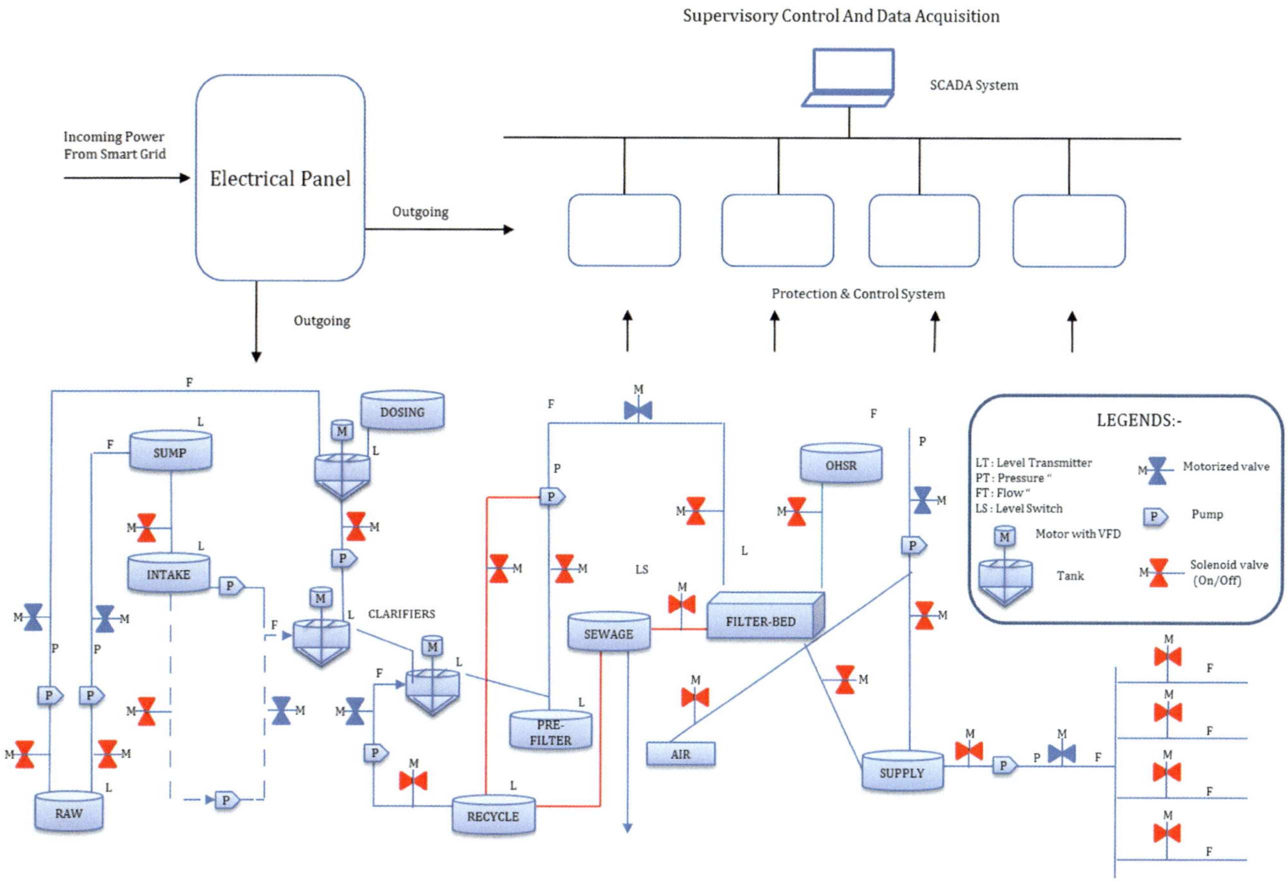

Fig. 1.11 The proposed architecture of WTP

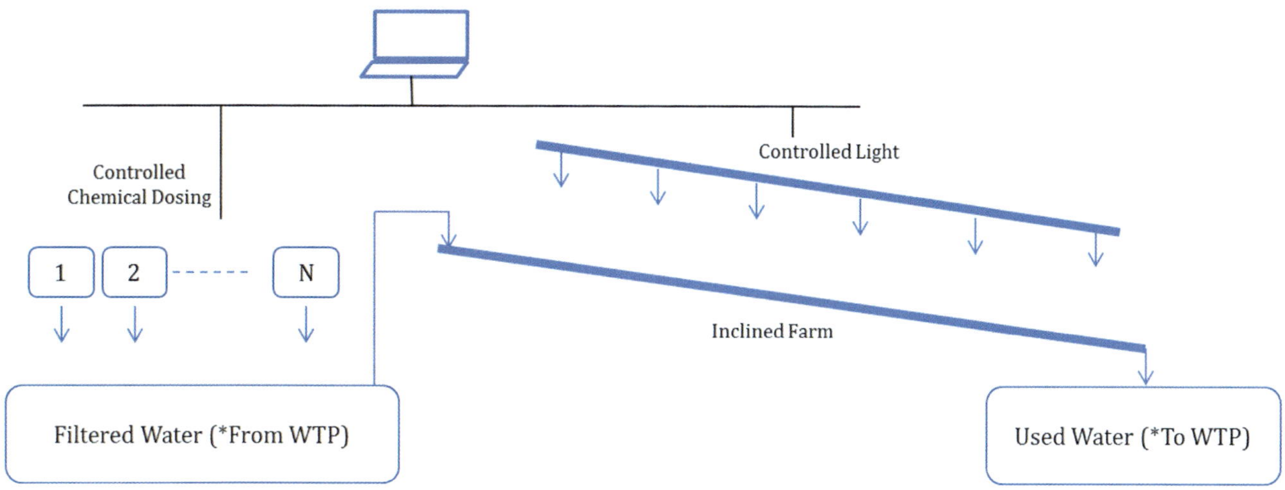

Fig. 1.12 Proposed architecture of smart farms

Smart Water Grid

The treatment for drinking water production involves the removal of contaminants from raw water to produce water that is pure enough for consumption without any short-term or long-term risk of any adverse health effect. The smart water-management systems accomplish these goals by increasing network visibility, facilitating predictive maintenance, and ensuring faster response times for events such as leaks, bursts, operational failures, quality incidents, and changes in water pressure.

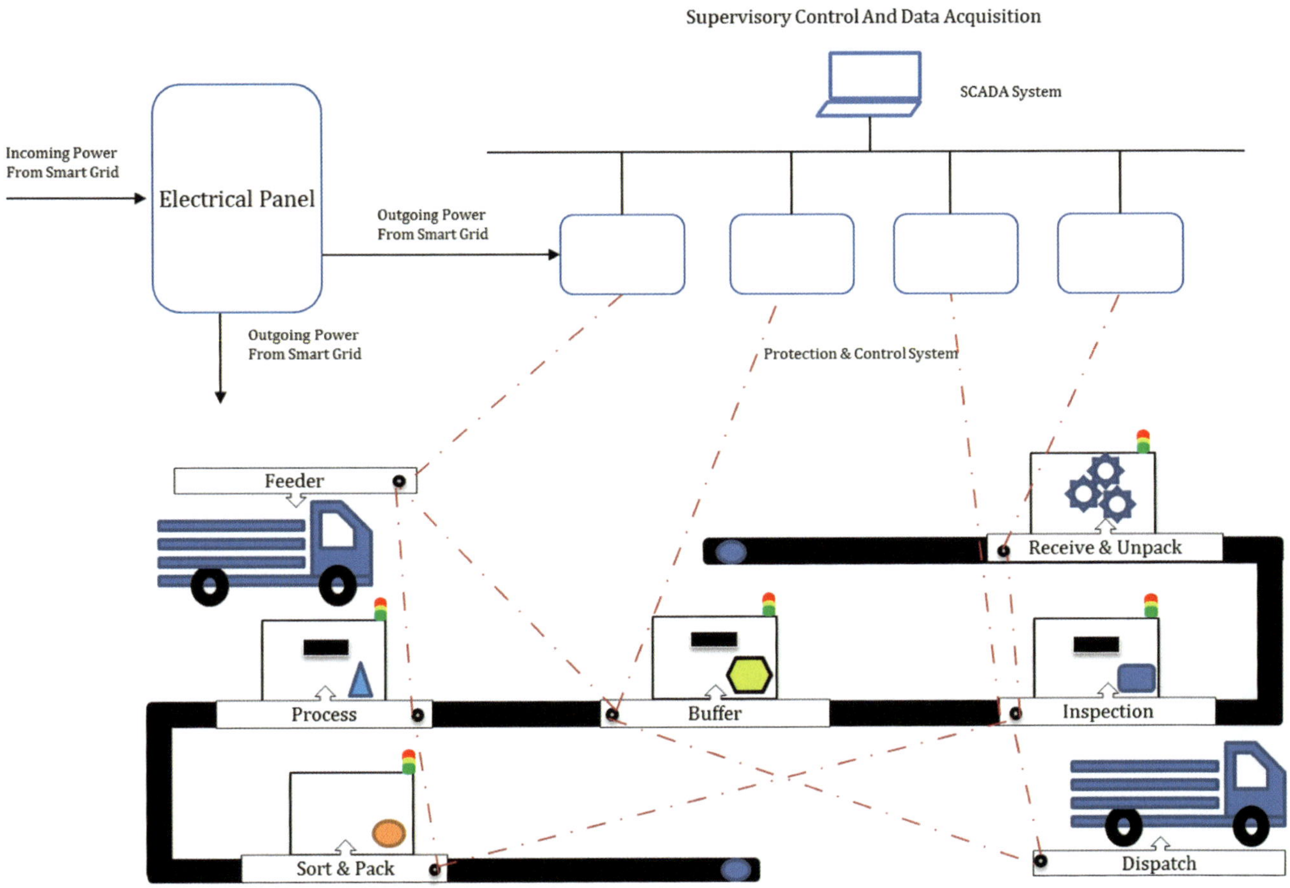

Fig. 1.13 The proposed architecture of smart industry

Table 1.6 The stations in manufacturing automation test-bed

S. No.	Station	Description
1	Feeder	This station separates the components from the Stack Magazine and distribute the components using rotary pick-and-place module
2	Inspection	This station measures the height of the components received from its downstream station and transfers the correct and incorrect components to appropriate slides
3	Buffer	This station ensures steady flow of components to the processing station by allowing one component at a time for processing. It buffers up to five workpieces at a time and if the count exceeded five, then it communicates with the upstream stations to hold the supply of the job until there is a demand from the downstream station
4	Process	This station demonstrates drilling operation on a pneumatically driven rotary indexing table and transfers the workpiece via pick-and-place module to downstream station
5	Sorting	This station sorts the incoming workpiece based on color and material characteristics to appropriate slides

As depicted in Fig. 1.11, C3i Center has WTP (water treatment plant) having number of pumps, motorized valves and instruments for measuring pressure, level, flow, pH, etc., equipped with protection and control system. This WTP treated contaminated water and supplies treated water for further use.

Treated Water Utilization

C3i center will have farm servers as depicted in Fig. 1.12, which will act as a load for smart grid and inclined pipe with small plants will act as a load for water treatment plant. Used water will again be circulated for treatment and facilitation of users with closed-loop control system.

Smart Industry

The smart manufacturing industries are one of the biggest consumers of electricity. As depicted in Fig. 1.13, C3i center has an isolated test-bed replicating smart industry which includes major components like feeder, inspection, buffer, process, and sorting. Detailed description of these stations is mentioned in Table 1.6.

References

1. Abbas HA (2014) Future scada challenges and the promising solution: the agent-based scada. Int J Crit Infrastruct 10(3–4):307–333
2. Igure VM, Laughter SA, Williams RD (2006) Security issues in scada networks. Comput Secur 25(7):498–506
3. Alcaraz C, Fernandez G, Carvajal F (2012) Security aspects of scada and dcs environments. Critical infrastructure protection. Springer, Berlin, pp 120–149
4. Ogie RI (2017) Cyber security incidents on critical infrastructure and industrial networks. In: Proceedings of the 9th international conference on computer and automation engineering. ACM, pp 254–258
5. Cryptocurrency attack on waste water site, USA. https://www.theregister.co.uk/2018/02/08/scada_hackers_cryptocurrencies/
6. Caldwell T (2018) Plugging it/ot vulnerabilities-part 1. Netw Secur 2018(8):9–14
7. Notpetya multiple targets, Ukraine & Germany ransomware. https://www.wired.com/story/notpetya-cyberattack-ukraine-russia-code-crashed-the-world/
8. Wannacry infection suspected in dacia production in Romania. http://business-review.eu/news/dacia-production-in-romania-partially-crippled-by-cyber-attack-wannacry-infection-suspected-137678
9. Crashoverride the malware that took down a power grid. https://www.wired.com/story/crash-override-malware/
10. Sfg malware discovered in European energy company. https://www.scmagazineuk.com/sfg-malware-discovered-european-energy-company/article/1476686
11. U.S. indicts 7 Iranians in cyberattacks on banks and a dam. https://www.nytimes.com/2016/03/25/world/middleeast/us-indicts-iranians-in-cyberattacks-on-banks-and-a-dam.html
12. Conficker kgg nuclear power facility, Germany worm. https://www.reuters.com/article/us-nuclearpower-cyber-germany/german-nuclear-plant-infected-with-computer-viruses-operator-says-idUSKCN0XN2OS
13. Cyber attack on U.S. water facility. https://constitution.com/recent-hacktivist-cyber-attack-us-water-utility-isnt-first-likely-last/
14. Encrypted firmware in plcs by a ransomware affected Ukraine power grid. https://en.wikipedia.org/wiki/2017_cyberattacks_on_Ukraine
15. Blackenergy malware caused blackout in Ukraine. https://en.wikipedia.org/wiki/December_2015_Ukraine_power_grid_cyberattack
16. Cyber attack on German steel mill. https://ics.sans.org/media/ICS-CPPE-case-Study-2-German-Steelworks_Facility.pdf
17. Korean energy and transportation targets attacked by oniondog. https://news.softpedia.com/news/korean-energy-and-transportation-targets-attacked-by-oniondog-apt-501534.shtml
18. America D, Cyber security in India: the nciipc road map
19. On a European programme for critical infrastructure protection (2005) EC, Brussels 17
20. Center for the protection of national infrastructure, UK. https://www.cpni.gov.uk/
21. Dayal A, Deng Y, Tbaileh A, Shukla S (2015) Vscada: a reconfigurable virtual scada test-bed for simulating power utility control center operations. In: 2015 IEEE power & energy society general meeting. IEEE, pp 1–5
22. Idaho national laboratory. https://www.inl.gov
23. Sandia national laboratories. https://www.sandia.gov/
24. Prabhakar T, Balaji V, Revathy K (2018) Mookit-a mooc platform for developing countries
25. Cyber security awareness week. https://security.cse.iitk.ac.in/node/523
26. Indian technical and economic cooperation programme (itec) ministry of external affairs, government of India. http://iitk.ac.in/oir/data/itec/Introduction-to-Cyber-Security.pdf
27. International society of automation. https://www.isa.org/
28. Mackiewicz RE (2006) Overview of IEC 61850 and benefits. In: 2006 IEEE PES Power systems conference and exposition, 2006. PSCE'06. IEEE, pp 623–630
29. Zhao Y, Shen ZJ (2003) Application of tcp/ip based iec60870-5-104 telecontrol protocol in power system [j]. Power Syst Technol 10:016
30. Huitsing P, Chandia R, Papa M, Shenoi S (2008) Attack taxonomies for the modbus protocols. Int J Crit Infrastruct Prot 1:37–44
31. Mills D et al. (1985) Network time protocol. Technical report, RFC 958, M/A-COM Linkabit
32. Why gis important smart cities. https://india.smartcitiescouncil.com/article/why-gis-important-smart-cities
33. Smart grid. https://www.smartgrid.gov/the_smart_grid/smart_grid.html
34. Smart home. https://www.smartgrid.gov/the_smart_grid/smart_home.html

The State of Android Security

2

Saurabh Kumar and Sandeep Kumar Shukla

Abstract

In today's world of devices, smart phones, tablets, and wearable devices, are widely used for communication, photography, entertainment, monitoring health status, and many more applications. Applications installed in the smartphones provide useful services, but they may maliciously send sensitive information to a remote location for various purposes. Due to the nature of open-source ecosystem, the usage of Android platform in mobile devices has grown significantly, and the security concerns have also increased manifold. Malware and software vulnerabilities issues violated end users' security and privacy. This article discusses and analyzes the Android architecture and platform vulnerabilities along with threat models, and malware analysis techniques followed by a few security challenges and future research directions.

Keywords

Android security • Mobile security • Data security • Privacy • Mobile malware • Malware analysis

2.1 Introduction

The Android Operating System is increasingly gaining popularity over the recent years. Android is preferred by users because of it free and open-source nature with support for a large number of applications. Developers also prefer Android over other commercial OS because of its open-source nature. The smartphone OS global market share presented by IDC shows that, in Quarter 1 of the year 2017, the total market share of Android was 85% [1]. With such rapid growth of Android OS, it has become the number one target for the malware authors.

Android is an open-source software stack based on Linux. It comprises of five layers, namely, Linux Kernel, hardware abstraction layer, Android Runtime, native libraries, application framework (Java API Framework), and application layer (default apps or third-party apps). Android Runtime is a virtual machine similar to JVM that executes/runs an Android application. The difference between Android virtual machine (ART/DVM) and JVM is that JVM is stack-based virtual machine whereas ART/DVM is register-based. Because of this, ART is faster than the JVM. Android is using two types of virtual machines, one is the Dalvik virtual machine DVM) [2,3], and the other one is ART (Android Runtime) [4]. The Dalvik virtual machine works on the principle of just-in-time (JIT) whereas ART works on the principle of ahead-of-time (AOT). Before Android version 5.0, DVM was the default runtime. From the Android version $\geq$5.0, Dalvik is entirely replaced by the ART.

Android security plays an important role in the Android ecosystem. In general terms, Android ecosystem refers to the interdependence between app users, app developers, and equipment makers. The ecosystem involves a number of actors each having their rights and duties: Platform developers are responsible for decisions regarding system and security, and device vendors are responsible for customizing the AOSP as per needs. App developers make application (apps) using different APIs built by library providers and tools offered by toolchain providers. App publishers are responsible for publishing the apps. Digital markets distribute apps from developers to the consumers in the ecosystem. All these actors have interdependency. Security impact on any one of the actors affects other actors as well. For example, if security in the case of OS developer is compromised, then it also affects the security of library provider, s/w developer, toolchain provider, s/w publisher, end user, etc. [5].

There is a need for improvement in the security of the actors in the ecosystem. An app should be vetted and verified by the App market before it can be published on the mar-

S. Kumar · S. K. Shukla (✉)
Indian Institute of Technology Kanpur, Kanpur, India
e-mail: sandeeps@cse.iitk.ac.in

S. Kumar
e-mail: skmtr@cse.iitk.ac.in

S. K. Shukla and M. Agrawal (eds.), *Cyber Security in India*, IITK Directions 4,
https://doi.org/10.1007/978-981-15-1675-7_2

ket. Similarly, toolkit providers can implement a mechanism which flags possible security issues to the developer [5].

For the last few years, malware has become a serious threat to Android users. The progress in Mobile security is not as fast as technology adoption in the smartphone, making it vulnerable to attacks. Additionally, due to the number of growing apps, hackers have found new ways to install malware straight to the devices. These are difficult to detect by traditional malware analysis techniques. Data from GDATA analytics shows, more than 3 billion new Android malware samples were found in the year of 2017, out of which 74,065 new malware were identified in the fourth quarter of 2017 [6]. This accounts for an average of 8225 new Android malware samples per day as per GDATA report.

The article is organized into five sections. Section 2.2 addresses the security architecture of the Android. Section 2.3 lists the security issues of Android in terms of vulnerabilities due to the architecture flaw, implementation issue, and hardware design issue. Section 2.4 describes the existing techniques used for malware analysis and their limitations. Section 2.5 concludes this article with the directions for future work.

2.2 The Security Architecture of Android

In this section, we look into the security features provided by the Android OS regarding Application Isolation (Sandbox) and the concept of permissions and least privilege.

2.2.1 Application Sandbox

The Application Sandbox governs an application's access right to the system resources. In Android, each application is isolated in its own environment. Android application can access only its own resources. An application needs proper access rights to access other sensitive resources. In Android, access rights are provided by permission, which needs to be declared in the Android manifest file.

To run each application in the isolated environment, Linux User ID model is used. Each application is run as a separate user and assigned a unique user id. Whenever an app is ready for execution, a new virtual machine (ART/DVM) is forked, and user id is assigned to the virtual machine. Figure 8.1 is showing the Android sandbox architecture.

In Fig. 2.1, App code (Classes.dex) file is written in Java (now Kotlin is also supported) and converted into the dex bytecode, which is then executed by the Android virtual machine (AVM). If an application contains native code, this code lies outside the virtual machine boundary. To use functionality available in native code, the AVM takes help of Java Native

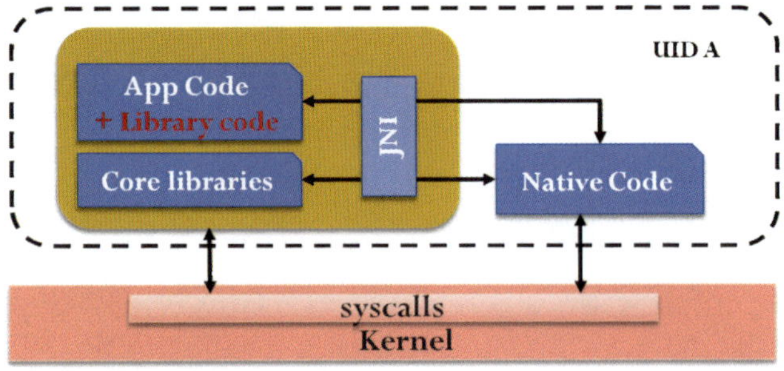

Fig. 2.1 Application sandboxing in Android

Interface (JNI). JNI is the bridge between code written in Java/Kotlin and native code.

2.2.2 Permissions

To accesses other resources which do not belong to an application is provided on the basis of permissions. In Android, each permission has a protection level, which describes a protocol for granting permission to the app. Permissions protection level is classified into four categories as shown in Table 2.1 [7].

2.3 Android Vulnerability and Advanced Threats

In the previous section, we have seen the Android security model, the features provided by the Android for security. This section discusses the vulnerability in Android system which can be exploited by an attacker for his/her malicious purpose. We have classified these vulnerabilities in three categories, namely, architectural, software vulnerability, and hardware vulnerability. First, we discuss these vulnerabilities which can be exploited by a malware. After that, we look at advanced techniques used by malware.

2.3.1 Architectural Vulnerability

The security issues related to the Android design are classified as architecture vulnerability. Example of such a vulnerability is collusion attack. In the collusion attack, two malicious apps can collude their permissions to gain extra privileges. Figure 2.2 is showing collusion attack [8] where two malicious apps A and B are involved. App A has Internet permission but not contacts read permission.

Similarly, App B has contact read permission but not Internet permission for sending these contact to the attacker's server. Both apps can merge their permission to send contact information to a server over the Internet. This permission

Table 2.1 Permission's protection levels

Protection level	Description
Normal	These permissions are granted to an app at install time and are protecting such resources which are not sensitive in terms of privacy issues
Dangerous	These permissions are related to the resources which are directly or indirectly violating the user privacy such as GPS, SMS, etc. Below the Android version 6.0, permissions are granted at install time, but from the Android 6.0 onward, these permissions can be adjusted by a user after app installation
Signature	These permissions are granted to only those apps which were signed by the same certificate
SignatureOrSystem	These permissions are similar to the signature protection level, but the app signing certificate is same as the certificate used to sign Android OS image

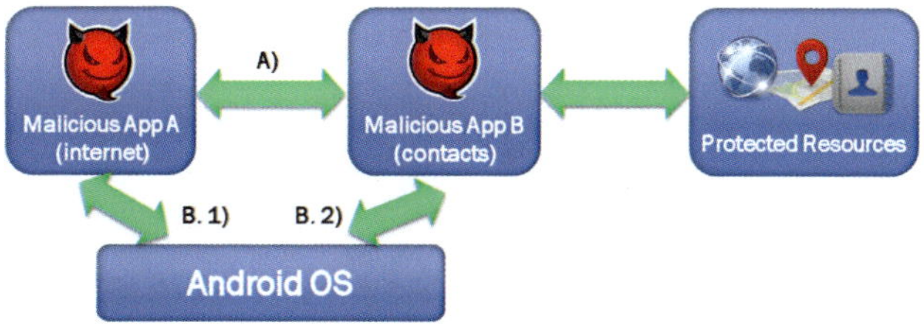

Fig. 2.2 Collusion attack

merging could take place if two apps are signed by using the same certificate (developed by the same developer). In this case, Android can assign the same user ID for both Apps. As the Android design provides this feature as its design principle, an attacker can take advantage to mount collusion attacks.

2.3.2 Software Vulnerability

The vulnerabilities which are introduced in the system during the implementation of the OS, Customizing the Android OS by the device vendor, Systems app provided by the device vendors and the Apps designed by a third party. These are categorized as software vulnerabilities. Example of these vulnerabilities is confused deputy attack [8], Media Projection Vulnerability [9], etc.

Media projection is an API introduced in Android 5.0 and above versions. It allows the application to capture the screen of third-party or system apps without the users consent. From Android 5.0 to 7.1, the API does not ask for user permission to record audio or capture screen of other apps. By exploiting this issue, an attacker can silently capture the screen when a user is accessing WhatsApp or any other app.

In Confused Deputy attack, a malicious app can fool a privileged app to perform some task on behalf of the malicious app. The confused deputy app can be a system app developed by the device vendor or a third-party app [8]. Multiple studies [10, 11] show that several confused deputies are around in the system apps and Samsung device firmware, in which one is running with the system privilege providing the root shell service to any other app.

2.3.3 Hardware Vulnerability

This issue comes because of hardware design flaws. Recently, an attack mounted on the DRAM is called RAMpage (CVE-2018-9442). Using this technique, an unprivileged Android APP can get root privileges. All the devices from the Ice Cream Sandwich (Android Version 4.0) to now are susceptible to RAMpage attack. RAMpage attack is a variation of the Rowhammer memory attack which can take advantage of the design flaw in modern memory systems.

2.3.4 Advanced Threats

This section discusses some techniques used by malicious apps and the risks associated with third-party libraries.

- **Risk of Third-Party Libraries:** A third-party library has to be included in every app that wants to use the library. As shown in Fig. 8.1, a third-party library executes in the same context as the app in which it is included. Because of this, a library has access to the host app's local and external files/data [13, 14]. If a library is compromised or developed by an attacker, then the data of the corresponding app is at risk. The study [15] shows that an average of 13 libraries per app is used in the top 3000 apps on GooglePlay.
- **Dynamic Code Loading Techniques and Risk:** It has been seen that most of the malware use dynamic code loading technique so that they can bypass malware analysis techniques. Table 2.2 shows a technique used for dynamic code loading and risk associated with it [12].

Table 2.2 Dynamic code loading: techniques and risks [12]

Techniques	API	Risk	Code injection vector
Class loader	DexClassLoader	No checking of **integrity** or **authenticity**	Attacker can control loaded code
Package context	CreatePackageContext	**No verification:** App from same developer	Attacker can install app
Native code	Java Native Interface	**No restriction** on location	Manipulate the native code to inject code
APK installation	PackageManager	No custom **verification** mechanism	Attacker can trick app/user into installing a malicious APK

Table 2.3 Malware analysis: techniques and challenges [16]

Techniques	Challenges
Static	**1. Dynamically** loaded code
	2. Crypto API
	3. Java reflection
	4. Network-based activity
Dynamic	**1.** False positive (**anomaly-based**)
	2. Code coverage
	3. Slow down in the system
Hybrid	**1. Logic**-based triggers
	2. Obfuscation and **Reflection**

2.4 Malware Analysis: Techniques and Its Limitation

As statistics show that, Android is the main target for malware authors, so there is need of Android malware analysis technique. This section reviews the existing malware analysis techniques and their limitations. The techniques used for malware analysis techniques are classified into three categories, viz., static analysis, dynamic analysis, and hybrid analysis [16], which are described in the following subsections.

2.4.1 Static Analysis

Using static analysis techniques, we can analyze an application/program without executing. The technique can be applied on Android application's source code or Android binary (APK). The static analysis approaches used on Android Apps for malware analysis are:

- **Signature-Based:** The signature-based approach is generally used in the commercial antivirus products. This method creates a unique signature by extracting the semantic pattern from the application. This signature is matched against the signatures of known malware in a database. If the signature of an app is not available in the signature database, then this app is declared as a legitimate app. Otherwise it is flagged as malicious. The shortcoming of this approach is that it cannot detect a malware whose signature is not in the signature database.
- **Permission-Based:** This technique takes advantage of the permissions used by an App. In this technique, rules are framed based on the set of permissions an app requests to detect a malicious App. It may be possible that the same set of permissions are required by a legitimate App, but this may lead to false positives.
- **Dalvik Bytecode-Based:** In Dalvik bytecode-based technique, the analysis is performed on the Dalvik bytecode to analyze the app behavior. To detect the malicious functionality performed by an App, control, and data flow analysis is carried out on the Dalvik bytecode.

2.4.2 Dynamic Analysis

In dynamic analysis, the behavior of an application is analyzed during its execution in real time. This method can detect malicious behaviors that are undetected by a static analysis technique. This technique suffers from the problem of code coverage, in which it may be possible that some code section may not execute during the analysis. The different techniques used in the dynamic analysis are:

- **Anomaly Detection:** This technique monitors the behavior of the system regularly. When the system deviates from

Table 2.4 Malware analysis frameworks [17]

Framework	Method	Limitation
Aurasium [Xu et al. 2012]	Dynamic + detect API misuse	Native code, Java reflection
DroidScope [Yan and yin 2012]	Dynamic	Limited code coverage
I-ARM-DROID [Davis et al. 2012]	Statically add stubs to use correct perm./API	Native code
SmartDroid [Zheng et al. 2012]	Statically find activity path + dynamic to find triggers	Native code
ContentScope [Yajin Zhou 2013]	Static path-sensitive data-flw + dynamic execution confirmation + classify leak/pollution	False positives (static and start errors), manual classification
Droid Analytics [Zheng et al. 2013]	Multilevel (method, class, payload) op-code signatures (static)	Logic obfuscation
VetDroid [Zhang et al. 2013]	Dynamic permission usage + reconstruct vine-grained actions	Native code
RiskMon [Jing et al. 2014]	Dynamic + machine learning + API monitor	Colluding apps
DroidSafe [Gordan et al. 2015]	Static information flow + hooks + calls that start activity	Dynamically loaded code

its regular behavior, it generates an alarm indicating an anomaly detection.

- **Taint Analysis:** In dynamic taint analysis, information flow tracking is performed during the execution of an application. In this technique, sensitive information is marked as the source, and the methods which may send this information outside the system are marked as the sink. When the system detects that a source is encountered at a sink, then an alarm is generated indicating possible leak of the sensitive information.

2.4.3 Hybrid Analysis

The hybrid analysis takes advantage of the static and dynamic analysis to monitor the malicious behavior of the application. In this technique, static and dynamic analysis approaches are combined to overcome the limitations of each other.

The challenges of malware analysis techniques are given in Table 2.3, and some malware analysis frameworks developed based on these techniques are shown in Table 2.4 with their limitations.

2.5 Conclusion

This article briefly outlines the security architecture of the Android and also discusses its vulnerabilities. GDATA statistics shows that around 343 new Android malware came every hour in 2017. This indicates that Android security mechanism is not up to the mark, and more fine-grained applications analysis techniques are needed so that a malicious app can be filtered out before publishing. To tackle this issue, a holistic system analysis approach is needed instead of a single application analysis. Not only this, we require to analyze dynamically loaded code which does not come bundled with an application, code written in the different form and different languages such as obfuscated code, native code, code written in Java/Kotlin, etc. As Android has completely shifted from DVM to ART, most of the past dynamic analysis work are on DVM. Therefore, there is an urgent need for dynamic analysis of apps executing in the ART environment.

References

1. Idc (2007) Worldwide smartphone os market share. https://www.idc.com/promo/smartphone-market-share/os
2. Ehringer D (2010) The Dalvik virtual machine architecture. Technical report, vol 4, no 8
3. Oh HS, Kim BJ, Choi HK, Moon SM (2012) Evaluation of android Dalvik virtual machine. In: Proceedings of the 10th international workshop on java technologies for real-time and embedded systems. ACM, pp 115–124
4. Idc: Worldwide smartphone os market share. https://source.android.com/devices/tech/dalvik/
5. Acar Y, Backes M, Bugiel S, Fahl S, McDaniel P, Smith M (2016) Sok: Lessons learned from android security research for appified software platforms. In: 2016 IEEE symposium on security and privacy (SP). IEEE, pp 433–451
6. G. data (2018) New android malware samples. https://www.gdatasoftware.com/blog/2018/02/30491-some-343-new-android-malware-samples-every-hour-in-2017
7. Android developers: permissions and description. https://developer.android.com/guide/topics/manifest/permission-element
8. Karthick S, Binu S (2017) Android security issues and solutions. In: 2017 international conference on innovative mechanisms for industry applications (ICIMIA). IEEE, pp 686–689
9. Hacking-news: android issue allows attackers to capture screen and record audio on 77% of all devices. https://latesthackingnews.com/2017/11/20/android-issue-allows-attackers-to-capture-screen-and-record-audio-on-77-of-all-devices
10. Android oem's applications (in)security and backdoors without permission. https://www.sh4ka.fr/Android_OEM_applications_insecurity_and_backdoors_without_permission.pdf

11. Felt AP, Wang HJ, Moshchuk A, Hanna S, Chin E (2011) Permission re-delegation: attacks and defenses. In: USENIX security symposium, vol 30, p 88
12. Poeplau S, Fratantonio Y, Bianchi A, Kruegel C, Vigna G (2014) Execute this! analyzing unsafe and malicious dynamic code loading in android applications. In: NDSS, vol 14, pp 23–26
13. Demetriou S, Merrill W, Yang W, Zhang A, Gunter CA (2016) Free for all! assessing user data exposure to advertising libraries on android. In: NDSS
14. Son S, Kim D, Shmatikov V (2016) What mobile ads know about mobile users. In: NDSS
15. Backes M, Bugiel S, Derr E (2016) Reliable third-party library detection in android and its security applications. In: Proceedings of the 2016 ACM SIGSAC conference on computer and communications security. ACM, pp 356–367
16. Tam K, Feizollah A, Anuar NB, Salleh R, Cavallaro L (2017) The evolution of android malware and android analysis techniques. ACM Comput Surv (CSUR) 49(4):76
17. Sadeghi A, Bagheri H, Garcia J, Malek S (2017) A taxonomy and qualitative comparison of program analysis techniques for security assessment of android software. IEEE Trans Softw Eng 43(6):492–530

3 Blockchain and Its Application in Cybersecurity

Shubham Sahai Srivastava, Ras Dwivedi, Abhishek Gunda, Devendra K. Meena, Rohit Negi, Nilesh Vasita and Ajay Singh

Abstract

Blockchains, in the past decade, have gathered a lot of attraction due to its potential to disrupt the way in which trust in the digital world is generated. Blockchain, the technology powering cryptocurrency worth billions of dollars, is tamper-resistant, immutable ledger that is maintained in a decentralized manner by the miners who do not trust each other. This characteristic of blockchain to induce trust in the untrusted environment, by establishing a commonly agreed upon (among the members of the blockchain network) shared ledger, can be used to solve problems in domains such as finance, supply chain, health care, IoT, etc. In this chapter, we will introduce blockchain and present its applications in cybersecurity.

Keywords

Cybersecurity • Blockchain • Cryptocurrency • Insider threat • Data integrity • Zero-knowledge proof

S. Sahai Srivastava · R. Dwivedi · A. Gunda · D. K. Meena · R. Negi (✉) · N. Vasita · A. Singh
Indian Institute of Technology Kanpur, Kanpur, India
e-mail: rohit@cse.iitk.ac.in

S. Sahai Srivastava
e-mail: ssahai@cse.iitk.ac.in

R. Dwivedi
e-mail: dwivedi@cse.iitk.ac.in

A. Gunda
e-mail: abhigun@cse.iitk.ac.in

D. K. Meena
e-mail: devk@cse.iitk.ac.in

N. Vasita
e-mail: nilesh@cse.iitk.ac.in

A. Singh
e-mail: sajay@cse.iitk.ac.in

3.1 Introduction

Blockchain, in the form of Bitcoin [1], was first introduced by an unknown person or group named Satoshi Nakamoto as a decentralized ledger, which is a data structure with a specific set of rules to add data, in chunks of blocks, that are chronologically arranged and immutable in the presence of adversaries. This is achieved by combining the properties of *cryptographic hash functions*, time stamping, and decentralization.

A *cryptographic hash function* can be thought of as a one-way function, that is, collision-resistant and non-invertible, i.e., given a hash value, it is tough to find its pre-image. Formally, a cryptographic hash function is a mapping $\{0, 1\}^* \longrightarrow \{0, 1\}^m$, such that it satisfies the following properties:

- *Deterministic*: Given x, $H(x)$ is unique.
- *Easy*: Given x, calculating $H(x)$ is computationally easy.
- *Non-invertible*: Given $H(x)$, it is computationally infeasible to find x' such that $H(x') = H(x)$.
- *Avalanche Effect*: A small change in x causes a drastic change in $H(x)$ and $H(x)$ is uniformly distributed.
- *Collision-Resistant*: It is computationally infeasible to find x_1 and x_2 such that $H(x_1) = H(x_2)$.

All the transactions in a given time frame are collected, ordered, time-stamped, and hashed to form a block. All the blocks, except the genesis block, also contain the hash of the previous block, thereby forming cryptographic linkages. It is through this cryptographic linkage that data is organized chronologically. At this point, it would be appropriate to draw an analogy between a book and a Blockchain data structure (Table 3.1).

Decentralization is the basis for the security of the blockchain. Arranging data in this fashion and linking them via hashes would not be fruitful or secure in case of centralized architecture, because calculating hash is easy

S. K. Shukla and M. Agrawal (eds.), *Cyber Security in India*, IITK Directions 4,
https://doi.org/10.1007/978-981-15-1675-7_3

Table 3.1 Analogy between a book and blockchain

Book	Blockchain
Data once written cannot be erased	Data once entered cannot be deleted
Pages in book could only be written in one direction	Blocks can only be added and not removed
All pages are numbered	All blocks are time-stamped and hashed
Number leads to arrangement of pages	Time stamp and cryptographic hash linkage lead to ordering of the blocks

and any such centralized database could be modified very easily. It is the decentralization along with the security of the cryptographic hashes that makes blockchain a trusted database, resistant to tampering.

Creation of a decentralized ledger means giving block-creating permission to every node (authorized node in the case of permissioned blockchain), and a fraction of those nodes could be malicious. So, one should be able to reach an agreement in the presence of these malicious nodes. There are various mechanisms for achieving the consensus. Cryptocurrencies like Bitcoin uses proof of work [1]. The essence of proof of work is that to add a block to the blockchain, one must present a proof of solving a computationally difficult cryptographic puzzle, the solution of which could be verified easily. Blockchain like Algorand [2] uses Byzantine Agreement protocol which works only if at least two-third of all the nodes are honest. This consensus algorithm of Algorand is computationally cheaper than the proof of work, but its requirement for fraction of node being honest is greater.

Blockchain became important because one can create a distributed database or ledger which could be verified just by checking the hash of the last few blocks, and not the entire data. Nobody has the power to tamper this database, even if they have read–write access, because that would imply breaking down *cryptographic hash function* which is computationally infeasible.

A distributed secure database, like blockchain, has its own drawbacks. Storage costs are increased because data has to be replicated across several nodes. Network load is also increased because data has to be transmitted to every node, not only for storage, but to reach consensus every node must have data and must share their views with other nodes.

In the following sections, we would present a brief evolution of the blockchains, their uses, and some of the theoretical and practical attacks.

3.2 Evolution of Blockchain

Blockchain as a technology has come a long way since its introduction in 2009 as Bitcoin. With the initial aim to solve the double spending problem and power a decentralized cryptocurrency, today as a technology it packs in many more capabilities such as supporting execution of trusted code via smart contracts, having faster consensus algorithms, associating every user (peer) with unique identities (permissioned blockchains), etc. These developments in the blockchain can be phased as Blockchain 1.0, 2.0, and 3.0. These phases in the blockchain development mark the distinct boundaries, where not only the capabilities of blockchain as a technology has increased but also new use cases have emerged with the potential of creating a revolution. In the rest of this section, we present these developments and capabilities.

3.2.1 Blockchain 1.0

Blockchain was introduced with the whitepaper of Bitcoin in 2009, which enabled the operation of a decentralized digital currency that can be operated on a trustless network, not governed by a set of individuals, organization, or governments. Anyone interested can become the part of the network and see what all transactions are happening across the globe, which gets appended in a single ledger. Furthermore, anyone is allowed to become the "maintainer" of the blockchain network, with the responsibility to check if the transactions happening are legitimate and are following the rules of the financial transactions supported by the corresponding cryptocurrency. These "maintainers" can select the legitimate transactions and combine them in a block and add them to the blockchain with the help of the consensus algorithm introduced: proof of work. These maintainers of the blockchain are known as miners, as in the process of creating these blocks, they get block rewards (currently, 12.5 Bitcoins get halved every 4 years), which is also the only method in which Bitcoins are created (mined). The technology was appreciated by many, and gave rise to a number of cryptocurrencies on similar lines to Bitcoin, with minor differences in their protocols, having the market cap of several hundred billion dollars.

Bitcoin blockchain supports a scripting language, often referred to as Bitcoin Script, which can be used to define the parameters that are needed to spend transferred Bitcoins. Bitcoin scripts are not Turing complete and do not support loops. Although they are sufficient to handle the parameters required for defining the rules for spending the Bitcoins, they cannot support general computations. It was soon realized that blockchain has the potential to go beyond the digital currencies and could solve exciting problems in the digital world. But, to achieve this, it was necessary to add more capabilities to the vanilla blockchain introduced by Bitcoin. These requirements lead to the next phase of the blockchain and the development of smart contracts.

3.2.2 Blockchain 2.0

In late 2013, Vitalik Buterin, a cryptocurrency researcher and programmer proposed Ethereum to realize this vision of blockchain going beyond cryptocurrency. With the launch of Ethereum, a blockchain that went live in July 2015, not only a new cryptocurrency (Ether) was released, but also a new revolution in the world of blockchain began. Compared to Bitcoin, in Ethereum, average time to generate a new block in the blockchain (block time) was reduced from 10 min to 15 s. For consensus, Ethereum uses a refined proof-of-work algorithm (Ethash), which reduces the advantages of ASICs in mining. Also, the block reward for mining the block in Ethereum is consistent (contrary to Bitcoin, which gets halved every 4 years), and also has a lower transaction fee as compared to Bitcoin.

Among the abovementioned improvements over the Bitcoin blockchain, Ethereum also supports scripting language much more advanced than Bitcoin script. It provided support for running code on the blockchain, called smart contracts, which can be written in a Turing complete programming language, Solidity. These capabilities on the Ethereum blockchain enable the community to develop decentralized applications or DApps, which can be thought of as open-source, decentralized applications that follow the protocol for execution across the distributed network. DApps, sometimes also referred to as DAOs (Decentralized Autonomous Organizations), became successful in raising millions using this new unregulated venture capital route also known as Initial Coin Offerings (ICO).

3.2.3 Blockchain 3.0

Ethereum laid the foundation of trusted computing and auditability with the help of smart contracts, but it turns out, still a lot more could be achieved. Due to the consumption of a large amount of computational resources and energy by proof of work, the reliance of the blockchains on PoW for consensus was the major bottleneck for practical applications. Another major concern was privacy; any transaction taking place on the blockchain was publicly visible to all on the blockchain ledger. This might not be desirable in a lot of practical applications.

Hence, the need of the hour was to overcome these shortcomings of blockchains, to have a more efficient consensus mechanism and enhanced privacy in the blockchain. Consequently, this phase of blockchain evolution has seen a lot of developments to improve upon these capabilities of blockchain and make them more scalable for practical applications and several different blockchain platforms came into existence. Some of them include IOTA (Blockchain platform based on "Tangle" and optimized of IoT devices), Stellar, Ripple, Hyperledger Fabric (Permissioned Blockchain), ZCash (supporting privacy based on Zero-Knowledge proofs), etc.

3.3 Security of Blockchains

Blockchains are considered secure by design, and any such secure system must meet fundamental quality attributes pertaining to security like CIA. CIA stands for confidentiality, integrity, and availability. In the rest of this section, we analyze blockchain from the perspective of confidentiality, integrity, and availability.

3.3.1 Confidentiality: Who Can See the Data

Blockchain is a shared ledger, and hence every entry is available to every participating node. Blockchain like Bitcoin claims pseudonymity as their real identity of the node is not available, and only its public key is visible. But there are methods to link public key to real-world identity. Confidentiality in transactions is not the part of the blockchain architecture, but it could be achieved by using an extra layer of cryptographic protocol over blockchain. Zcash is one such cryptocurrency that uses zk-SNARKS to achieve confidentiality. We discuss more about confidentiality in Sect. 3.4.1.

3.3.2 Integrity: The Accuracy and Consistency of Data over Its Entire Life Cycle

In blockchain, blocks could only be added and not updated or deleted. Every block, except the genesis block, contains the hash of the previous block. So, if one tries to change data stored in a block, one would have to change the hash values of all subsequent blocks. For every block, there is consensus on the hash of the block. As a result, either an attacker would have to do a hash collision or he/she would have to change the hash of the latest block. So, as long as the hash of the last block has not changed, one could be sure about the integrity of the data.

3.3.3 Availability: Liveness Guarantee

Availability could be expressed as the degree to which a system is operational and accessible when required [3]. Blockchain is a distributed system, in which anybody could be a miner. A miner has the responsibility to collect all the transactions, order them, and create a block, which would then be accepted by other nodes. As long as there is even one single miner, blockchain exists, and hence blockchain is immune to DDoS attack because attacking miners, distributed all across the globe, on an Internet scale is largely infeasible.

3.4 Applications

3.4.1 Secure Decentralized Currency

Cryptocurrencies have gained a lot of popularity adhering to the secure modes of transactions. One such example is Bitcoin. Bitcoin is a decentralized digital currency without a central bank or single administrator and can be sent from user to user on the peer-to-peer Bitcoin network without the need for intermediaries. With Bitcoin, transactions are made pseudonyms, and payment transactions are recorded in a decentralized public ledger. Although bitcoin transactions are somewhat claimed to be anonymous, much information can be deduced from the public ledger [4], i.e., it may allow data miners to link individual transactions, identify related payments, and, otherwise, trace the activities of Bitcoin users.

Zerocoin protocol [5] tackles some of these privacy issues by unlinking transactions from the payment's origin. Yet it still reveals payment destinations and amounts and is limited in functionality. Now, we would discuss how zero-knowledge proofs are used by Zerocash (Zcash) [4] which can be used for making transactions secure.

Zcash relies on zk-SNARKs (zero-knowledge succinct non-interactive arguments of knowledge), a novel form of zero-knowledge cryptography. Transactions in Zcash can be fully encrypted on the blockchain, yet still be verified as valid under the network's consensus rules by using zk-SNARK proofs.

"Zero-knowledge" proofs allow one party (the prover) to prove to another (the verifier) that a statement is true, without revealing any information beyond the validity of the statement itself. For example, given the hash of a random number, the prover could convince the verifier that there indeed exists a number with this hash value, without revealing what it is. Zero-knowledge proofs can thus be used to validate transactions without revealing any information about the transaction (i.e., zero knowledge). Construction of such a proof is equivalent to proving a mathematical equation which satisfies the above conditions.

3.4.2 Smart Contracts

We can think of smart contracts [6] as a set of conditions written in an executable code which runs on top of blockchain and is used to carry out transactions between two or more untrusted parties without the involvement of a trusted third party. The trust on smart contracts comes from the fact that their code is tamper-resistant as they are deployed on the blockchain. If the parties want to edit the set of conditions then they have to create a new smart contract and again deploy it to the blockchain. Every smart contract is assigned a unique address on the blockchain. Different blockchain platforms like Etherium, Hyperledger, and NXT can be used to develop and deploy the smart contracts. Let us understand it better from a simple, e.g., suppose a company uses smart contracts on Etherium to distribute 20 ether among its 5 employees equally. The payroll manager will send 20 ether to the address of the smart contract. The addresses of employees account will be hardcoded in the smart contract's executable code with the condition of equal distribution. The code will get executed, and four ether will be sent to each employee's account. Smart contracts can be used to enforce the complex set of conditions and can be used to carry out different kinds of transactions, i.e., not only to transfer cryptocurrencies. Some other applications could be

1. Automation of insurance companies,
2. Distribution of salaries among employees, and
3. Decentralized social media network implementation.

3.4.3 Document Verification in KSI

Document verification is the procedure to assure the integrity of documents. The key issues in document verification are storage, retrieval, access to data, and most importantly avoiding the possibility of any manipulation of document's content. Reference [7] KSI is an infrastructure developed in Estonia and is being used to ensure networks, system, and data integrity, all while retaining 100% data privacy. KSI architecture [8] shown in Fig. 3.1 uses the concept of Merkle tree (binary tree with parent node containing the combined hash of its child nodes) comprising mainly of three components, namely, gateway layer, aggregation layer, and a core cluster.

It is a permissioned network where document's hash values are passed at the gateway layers. The gateways aggregate these hash values, and the top node of the Merkle tree created at each gateway is then passed to the aggregator one layer above. The root hash values from different gateways are aggregated again and sent again one layer above and eventually to the core cluster. The core computes a single current uppermost core hash value at the respective tree node at each calendar time (could be 1 s). This uppermost value is referred to here

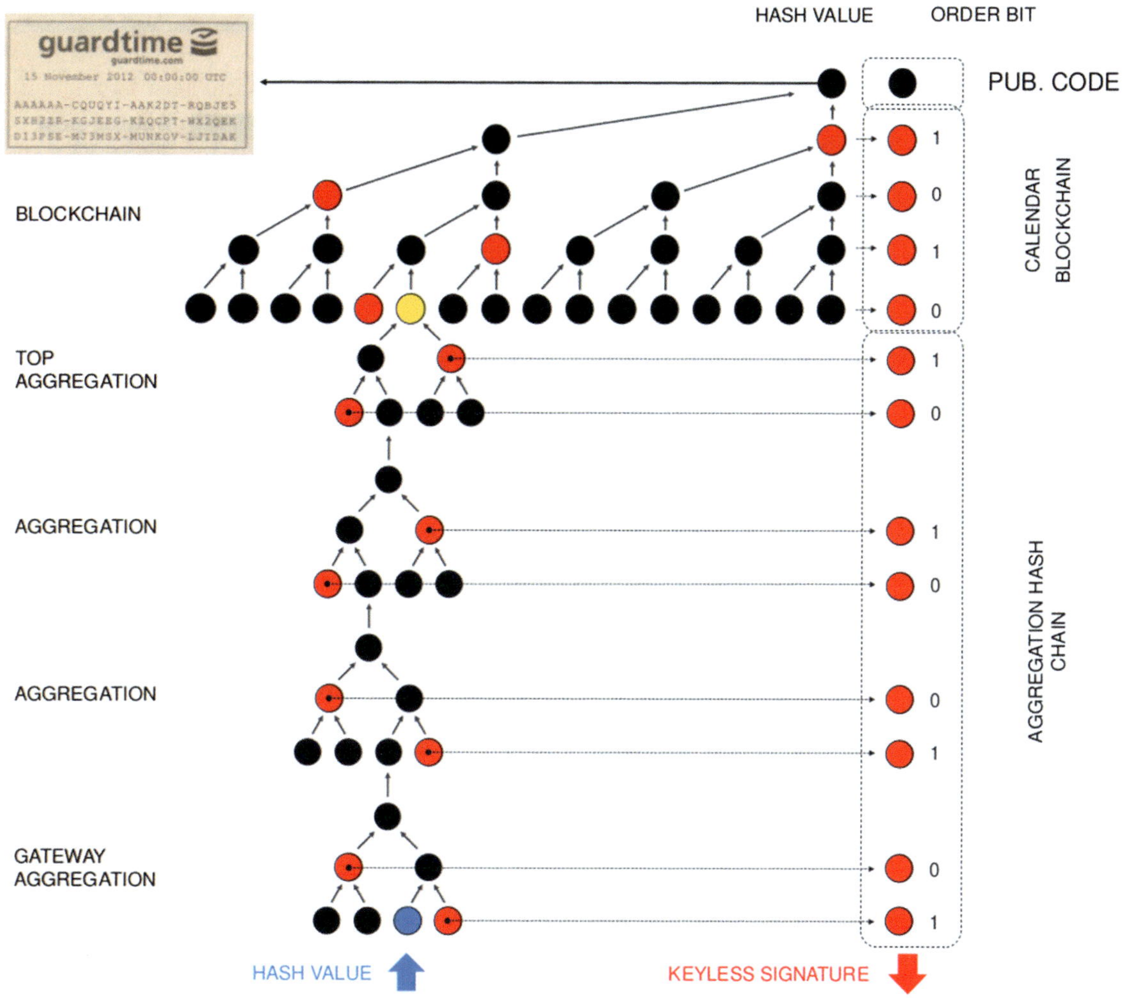

Fig. 3.1 KSI architecture

as the “calendar value”. A Merkle tree is again created on top of these calendar values collected over time, and the final root value is published in a newspaper. Once the calendar value is computed by the core cluster, it returns a token (or signature) back to the client, which are the hash values of the siblings that can be used to reconstruct the path to the published root value. The signature vector is shown in Fig. 3.1 where 0 represents left sibling and 1 represents the right sibling. There are two levels of verifications:

- **First level**: Calendar values are stored in a database at core cluster, and a copy of that database is also in sync at gateways. Hash values of siblings are enough to reconstruct the path and verify if the calendar value is same as the computed one.
- **Second level**: This level of verification is executed using Merkle tree that is created in calendar blockchain. It can be done only after the publishing time period is complete.

KSI is scale-free as it can sign and verify 10^{12} transactions/sec. Also, it is quantum immune as it does not rely on the PKI and uses only hashes. Currently, it is being used in Estonia for applications in government sectors like land registry, business registry, e-health, court cases, etc. There could be other use cases like securing SDNs, systems, or even websites to detect many cyberattacks.

3.4.4 Insider Threats

Insider threats are the threats posed by individuals who already have some privileges and are authorized to use the resources of the organization. These individuals may be the employees, business partners, or even system administrators. It is a well-known fact that most companies nowadays take utmost care to secure their digital assets and mitigate external attacks, all the while neglecting the insider threats. Insider attacks are a tougher problem in the information security to be dealt with, as a malicious insider is too well concealed to be detected by conventional methods. Many of the companies often lack the tools/infrastructure to deal with the risks relating to malicious insiders.

Many insider attacks are detectable by using proper logging mechanisms which are appropriately alienated from production systems and employees. But with all applications certainly relying on a centralized databases/architecture it is highly probable that any employee with administrative privileges can attack the system and even corrupt the logging mechanism making the central hub a single point of failure. This dependence of relying on centralized architecture is what blockchain technology aims to prevent and the storing of data in a decentralized and distributed manner and the immutable property of data certainly help to prevent from compromising large volumes of data.

3.4.5 Supply Chain Management

Nearly all of the world's leading logistics companies maintain a supply chain management software connecting manufacturing to shipping and RFID of products. Despite this huge investment, the visibility of products seems to be very limited due to interoperability and inconsistency over the enterprise boundaries.

We have already seen that the blockchain is immutable and the data on the blockchain ledger cannot be modified or deleted. So, every time a product changes hands, we can implicitly create a transaction to be added on the blockchain and thus leave a trail (of transactions) which are immutable, owing to the properties of the blockchain, thus creating a secure, permanent, and transparent historical record of the product supply chain, from manufacture to the final buyer.

The advantage of blockchain is that companies gain a real-time digital ledger of transactions and movements of products for all the participants across the enterprises which brings greater efficiency to the companies and also reduce the time delays and human errors that plague the transactional costs.

Advantages:

- Logistics: Recording every transaction between supply chain nodes leads to a better understanding of the statistics and can bring more in data analysis.
- Scalability: Virtually any number of participants can be added to the block and thus it has the potential to become the universal supply chain model.
- Payments: Most often the payment of the goods is delayed due to the validity of product delivery and this can be mitigated by the use of smart contracts, thus making the payments digital and seamless.
- Security: The transparency and immutability of ledger reduce the frauds and the need for audits.
- Transparency offers: Transparency offers proof about how goods were sourced and how they comply with regulations. The physical, financial, and digital information are brought together in one platform to reveal sources of value leakage from everyday inefficiencies.

3.4.6 Patch Management, Backing Up, and Restoration: Enforcing Policy

To keep infrastructure secure and less vulnerable to adversaries, we have to keep patching our cyber assets. We also need to minimize the shutdown time, manage the risk, and frequently back up our systems. However, before restoration to a past backup, the integrity of the backup should be verified.

The requirements of an operational system are quite different from that of information systems. Critical components of critical infrastructure like power, gas, oil, etc. include industrial control systems and protection systems. With the upcoming Industry 4.0 [9] and IT-OT convergence, plant floor operations are further integrated with the business environment. This change is exposing operational systems to the business network which means more adversaries. These fully automated, geographically distributed critical infrastructures are now becoming a lucrative target for cyberattacks. Therefore, keeping the systems patched is of utmost importance. Patching a critical infrastructure is itself a challenge because of Legacy hardware and software systems, requirement to minimize shutdown time, and the lack of timely awareness of new patch availability.

Critical infrastructures are obligated to keep running necessitating zero shutdown or negligible shutdown efficiently. During preventive maintenance, they can be patched, but one should keep the patch ready and tested, to avoid any undesired scenario. If the patch fails to satisfy functional requirement, then the system needs to be restored to a previous backup. Sometimes, it might be safer to move to an earlier stable backup than the latest one. One can ensure the integrity of the past backups by storing the hash values of the backups along with relevant meta-information on a blockchain. However, searching for a backup hash on a blockchain might be difficult because indexing like the relational database is not available in most blockchain platforms. Fortunately, a few blockchains dump the data into an external database with integrated indexing and searching capabilities [10].

In summary, keeping the operational platforms safe from known exploits, one must consider the following requirements:

- Taking backup regularly.
- Keeping integrity of the backup through tamper-resistant hashing.

- Keeping the hash values intact through the blockchain.
- Either applying a patch or restoring from backup.
- Patch verification through the digital certificate from the company issuing the patch.
- Verification of the integrity of the backup through searching hash values on the blockchain.

3.4.7 Blockchain-Based Authentication

With the help of blockchain, a security system used in an organization can leverage a distributed public-key infrastructure for authenticating devices and users. This security system provides each device with a specific SSL certificate instead of a password. Management of certificate data is carried out on the blockchain, and this makes it virtually impossible for attackers to utilize fake certificates.

3.5 Blockchain in IoT

The Internet of Things (IoT) is expanding at a fast pace and some reports [11] predict that IoT devices will grow to 26 billion by 2020, which are 30 times the estimated number of devices deployed in 2009 and is far more than the 7.3 billion smartphones, tablets, and PCs that are expected to be in use by 2020. To facilitate such a huge growth, it is necessary to build an IoT stack, standardize protocols, and create the proper layers for an architecture that will provide services to IoT devices. Currently, most IoT solutions rely on the centralized server–client paradigm for connecting to cloud servers through the Internet. As seen in the previous sections, using blockchain technologies one can track, coordinate, carry out transactions, and store information from a large number of devices, enabling the creation of applications that require no centralized cloud.

The critical challenges faced by IoT for its massive adoption [12] includes the following:

- Expensive deployment and maintenance costs of centralized clouds and server farms resulting in expensive IoT solutions.
- Remotely maintaining million of devices for regular software updates.
- Privacy and anonymity for an IoT user are not guaranteed. How can a user be assured that the data generated from the IoT devices is not being accessed by any unauthorized third party for surveillance and analysis?
- Lack of trust among the users is also fostered by closed-source code.

A common problem shared by IoT and cryptocurrencies is that there are many entities (nodes, gateways, users) that do not necessarily trust each other at the time of performing transactions.

3.5.1 Use of Blockchain in the IoT Space

The first step for talking about blockchains in the IoT paradigm is to figure out if blockchain is even needed for a given IoT application? One can think of leveraging blockchains for IoT applications if any of the following properties are required in the application: decentralization, P2P exchanges, payment system, public sequential transaction logging, robust-distributed system, and micro-transaction collection.

The use of blockchain for enhancing the experience of IoT applications is an active area of research. Some of the previous use cases of blockchain for IoT applications include [29], like sensing [13,14], data storage [15,16], identity management [17], time-stamping services [18], smart living applications [19], intelligent transportation systems [20], wearables [21], supply chain management [22], mobile crowd sensing [23], agricultural applications [24], smart cities [25], cyber law [26], and security in mission-critical scenarios [27]. Other researchers focused on managing IoT devices through a blockchain [28]. The healthcare industry also does not lag behind in leveraging the blockchain technology. Some of the published work in this space include the use of blockchain to verify data integrity and public accessibility to temperature records in the pharmaceutical supply chain [29], while others use it for clinical trials and precision medicine [30]. A smart healthcare system using blockchain is also proposed in [31]. The energy sector can also be benefited from the application of a blockchain to IoT or to the Internet of Energy (IoE) [9]–[32]. A blockchain-based system detailed in [33] allows IoT/IoE devices to pay each other for services without human intervention.

3.5.2 The Important Aspects of Blockchain-Based IoT Applications

Indicating only the positive sides of the blockchain would be imprudent to the challenges faced by the industry in adopting blockchain-based IoT applications. The current challenges can be categorized widely in the following categories:

- **Privacy**
  The transparent transactions fostered by blockchains are a challenge in terms of privacy, i.e., the data once added to the blockchain is added forever and is open in the public domain, which allows third parties to monitor and predict on people behavior or habits. However, note that in many IoT applications anonymity is not necessary, but

the privacy of the transactions is required in certain scenarios when the collected data may allow for monitoring and predicting people behavior or habits. One solution for increasing privacy is through the use of zero-knowledge proving techniques like the ones used by Zerocash [34] or Zcash [35]. Another possible solution for preserving privacy could be the use of homomorphic encryption [36,37]. Such a kind of encryption allows third-party IoT services to process a transaction without revealing the unencrypted data to those services.

- **Security**
  Conventionally, security of an information system is determined by its ability to fulfill the three requirements, viz., confidentiality, integrity, and availability. Confidentiality of data is already covered under the privacy aspect. With respect to the infrastructure that supports IoT applications, confidentiality is achieved as long as the centralized infrastructure is trusted and proves to be robust against various kinds of cloud computing attacks [38,39] and internal leaks. For data integrity, IoT applications rely on third parties whereas blockchain by its definition comes along with data integrity. The work in [40] proposes one such use case for integrity service framework for cloud-based IoT applications. Similarly, availability also by design is assured by blockchains. However, some attacks like the 51-percent attack can create serious availability issues for performing transactions which eventually affects the integrity of the data.
- **Energy efficiency**
  The major issue with the IoT end nodes is that they make use of resource-constrained hardware that is powered by batteries. Many blockchains, including bitcoin, use the very resource heavy PoW consensus algorithm for mining. Hence, the selection of a suitable consensus algorithm for mining in the IoT space becomes very important. Also, new hashing algorithms like Scrypt [41] or X11 [42] are faster and thus can reduce mining consumption.
- **Throughput and Latency**
  It is one of the major issues with the use of blockchain for IoT applications. The throughput of the popular blockchains, like that of bitcoin, is far less than its contemporaries.
- **Blockchain size, Bandwidth, and Infrastructure**
  The resource-constrained IoT end devices cannot store the whole blockchain as its size increases periodically. Hence, this calls in for a need of powerful nodes in the IoT hierarchy. Also, it should be noted that transaction and block size have to be scaled according to the bandwidth limitations of IoT networks: many small transactions would increase the energy consumption associated with communications, while a few large ones may involve big payloads that cannot be handled by some IoT devices.

3.6 Attacks on Blockchain-Based Systems

Blockchain has revived trust by the use of "Cryptography" and "Consensus Mechanism" which provides security, anonymity, and data integrity without the involvement of the third party. But there are still some limitations and technical challenges which makes blockchain prone to attack. Some of the possible threats and the attacks that happened in the past are mentioned below.

3.6.1 51% Attack

It is a type of attack on blockchain where a group of miners possesses more than 50% of the available mining power and thus have the power to (or at least attempt to) make a transaction get approved or not. Let us consider an example: let there be two miners A and B in a network with 60% and 40% of hash power, respectively. This hash power signifies that for every n blocks, $0.6n$ blocks are owned by A and $0.4n$ by B. In due course of time, A's and B's chains vary significantly with respect to the number of blocks. It is tough for B to take over the A's chain with less hash power. In blockchain, the longest chain is preferred, so B's chain will be discarded. Hence, if anyone possesses more than 51% hash rate then they will lead the chain with very high probability. The scenario is depicted in the figure below:

Some examples of the 51% attacks over the years are discussed below:

- Two Ethereum-based blockchains Krypton and Shift suffered from 51% attacks in August 2016.
- Bitcoin gold cryptocurrency suffered from 51% attacks in May 18, the malicious miner controls more than 51% of Bitcoin Gold's mining power and steals $18 million worth of Bitcoin Gold by double spending.

3.6.2 Mining Pool Attack

Mining pool comprises the set of miners who work together on a problem to increase the chance of winning the mining reward. There is a pool manager who forwards the unsolved work unit to the pool members. The miners generate *partial proofs of work* and *full proofs of work* and submit this to manager as contribution to there part. The manager then publishes the block to the network and distributes the earned reward within the pool as per the contribution made by each of the members. Now, when the miner tries to get more rewards then his contribution by some malicious act in the pool is called the mining pool attack. In this, the malicious miner tries to fool the manager by submitting the partial proofs of work

(If it is asked to provide the hash with 10 leading 0s and you are submitting the hash with 9 leading 0s) and try to mine the block secretly to gain block reward for himself.

3.6.3 Eclipse Attack

A malicious agent manipulates a victim peer, and finally it leads to the network partition between the victim and public network. The malicious agent makes a false view of blockchain to the victim and bound him to work on the chain on which he wants. With the help of this, the attacker can launch a 51% attack with less mining power.

The attacks mentioned above give the flavor of theoretical attacks possible on blockchains, but there are very prominent mitigation techniques that exist, making exploitation extremely difficult. Now, let us see some real-life attacks that took place in the recent past.

3.6.4 Practical Attacks [43]

- On August 15, 2010, an attacker was able to generate 184 billion Bitcoin transaction in a single block, exploiting an integer overflow vulnerability in the code.
- In January 2015, Bitstamp exchange was hit by malware and 19,000 Bitcoins were stolen. The attacker has distributed files containing malware by appealing to their personal histories and interests.
- On August 2, 2016, the attackers exploited a vulnerability with Bitfinex's multi-signature wallets used to store their customer's funds. A total of 120,000 Bitcoin were stolen from the wallets.
- In July 2017, $7 million was stolen from CoinDash an Israeli start-up that conducted an ICO. The attacker changed the Ethereum address posted on the ICO's website. This is the address used by the investors to exchange their Ether with CoinDash tokens.

3.7 Conclusion

In this chapter, we have introduced blockchain as a tamper-resistant distributed ledger. We have discussed about its security guarantees and shown that it has the potential to become the backbone of modern cybersecurity infrastructure. We have also presented some of the attacks on blockchain and shown that blockchain is only as safe as their implementation is. So, even if theoretical guarantees of blockchain are sound, one has to be sure about its implementation.

References

1. Nakamoto S (2008) Bitcoin: a peer-to-peer electronic cash system. http://bitcoin.org/bitcoin.pdf
2. Micali S (2016) ALGORAND: the efficient and democratic ledger. CoRR arXiv:abs/1607.01341
3. http://www.mit.jyu.fi/ope/kurssit/TIES462/Materiaalit/IEEE_SoftwareEngGlossary.pdf
4. Ben-Sasson E, Chiesa A, Garman C, Green M, Miers I, Tromer E, Virza M (2014) Zerocash: decentralized anonymous payments from bitcoin. In: 2014 IEEE symposium on security and privacy, pp 459–474
5. Miers I, Garman C, Green M, Rubin AD (2013) Zerocoin: anonymous distributed e-cash from bitcoin. In: 2013 IEEE symposium on security and privacy (SP). IEEE, pp 397–411
6. Blockchain based smart contracts: a systematic mapping study (2017)
7. Ksi us google patent 2017 (2017)
8. Advanced trust services facilitated by the industrial - scale blockchain technology (2015)
9. i4.0. https://en.wikipedia.org/wiki/Industry_4.0
10. Availability. https://www.linkedin.com/pulse/quality-attributes-blockchain-chayan-keshari
11. Forecast: the internet of things, worldwide, 2013. Gartner, Stamford (2013)
12. IBM: device democracy: saving the future of internet of things. IBM, New York (2015)
13. Wörner D, von Bomhard T (2014) When your sensor earns money: exchanging data for cash with bitcoin. In: Proceedings of the UbiComp Adjunct, Seattle, pp 295–298
14. Zhang Y, Wen J (2015) An iot electric business model based on the protocol of bitcoin. In: Proceedings of the 18th international conference on intelligence in next generation networks, Paris, France, pp 184–191
15. Wilkinson S et al Storj a peer-to-peer cloud storage network. https://storj.io/storj.pdf. Accessed 10 April 2018
16. Ateniese G, Goodrich MT, Lekakis V, Papamanthou C, Paraskevas E, Tamassia R (2017) Accountable storage. In: Proceedings of the international conference on applied cryptography and network security, Kanazawa, Japan, pp 623–644
17. Wilson D, Ateniese G (2015) From pretty good to great: enhancing pgp using bitcoin and the blockchain. In: Proceedings of the international conference on network and system security, New York, pp 368–375
18. Gippand B, Meuschke N, Gernandt A (2015) Decentralized trusted timestamping using the crypto currency bitcoin. In: Proceedings of the iConference Newport Beach, CA, USA, pp 1–5
19. Han D, Kim H, Jang J (2017) Blockchain based smart door lock system. In: Proceedings of the international conference on information and communication technology convergence (ICTC), Jeju Island, South Korea, pp 1165–1167
20. Lei A, Cruickshank H, Cao Y, Asuquo P, Ogah CPA, Sun Z (2017) Blockchain-based dynamic key management for heterogeneous intelligent transportation systems. IEEE Internet Things J 4(6):1832–1843
21. Siddiqi M, All ST, Sivaraman V (2017) Secure lightweight context-driven data logging for bodyworn sensing devices. In: Proceedings of the 5th international symposium on digital forensic and security (ISDFS), Tirgu Mures, Romania, pp 1–6
22. Kshetri N (2017) Can blockchain strengthen the internet of things? IT Prof 19(4):68–72
23. Tanas C, Delgado-Segura S, Herrera-Joancomartí J (2016) An integrated reward and reputation mechanism for mcs preserving users' privacy. In: Proceedings of the 10th international workshop on data privacy management, and security assurance, vol 9481. Springer, New York, pp 83–89

24. Kshetri N (2016) An agri-food supply chain traceability system for china based on RFID & blockchain technology. In: Proceedings of the 13th international conference on service systems and service management (ICSSSM), Kunming, China, pp 1–6
25. Biswas K, Muthukkumarasamy V (2016) Securing smart cities using blockchain technology. In: Proceedings of the IEEE 14th international conference on smart city, Sydney, NSW, Australia, pp 1392–1393
26. Wright A, De Filippi P (2015) Decentralized blockchain technology and the rise of lex cryptographia. https://ssrn.com/abstract=2580664. Accessed 10 April 2018
27. Kshetri N (2017) Blockchain's roles in strengthening cybersecurity and protecting privacy. Telecommun Policy 41(10):1027–1038
28. Huh S, Cho S, Kim S (2017) Managing IOT devices using blockchain platform. In: Proceedings of the 19th international conference on advanced communication technology, (ICACT), Bongpyeong, South Korea pp. 464–467 (Feb. 2017)
29. Bocek T, Rodrigues BB, Strasser T, Stiller B (2017) Blockchains everywhere-a use-case of blockchains in the pharma supply-chain. In: Proceedings of the IFIP/IEEE symposium on integrated network and service management (IM), Lisbon, Portugal, pp 772–777
30. Shae Z, Tsai JJP (2017) On the design of a blockchain platform for clinical trial and precision medicine. Proceedings of the IEEE 37th international conference on distributed computing systems (ICDCS), Atlanta, pp 1972–1980
31. Salahuddin MA, Al-Fuqaha A, Guizani M, Shuaib K, Sallabi F (2017) Softwarization of internet of things infrastructure for secure and smart healthcare. Computer 50(7):74–79
32. Fernández-Caramés TM (2015) An intelligent power outlet system for the smart home of the internet of things. Int J Distrib Sens Netw 11:214805
33. Lundqvist T, de Blanche A, Andersson HRH (2017) Thing-to-thing electricity micro payments using blockchain technology. In: Proceedings of the global internet things summit (GIoTS), Geneva, Switzerland, pp 1–6
34. Zerocash. http://zerocash-project.org. Accessed 10 April 2018
35. Zcash. https://Z.cash. Accessed 10 April 2018
36. Moore C, O'Neill M, O'Sullivan E, Doroz Y, Sunar B (2014) Practical homomorphic encryption: a survey. In: Proceedings of the IEEE international symposium on circuits and systems (ISCAS), Melbourne, VIC, Australia, pp 2792–2795
37. Hayouni H, Hamdi M (2016) Secure data aggregation with homomorphic primitives in wireless sensor networks: a critical survey and open research issues. In: Proceedings of the IEEE 13th international conference on network, sensor, control (ICNSC), Mexico City, Mexico
38. Jabir RM, Khanji SIR, Ahmad LA, Alfandi O, Said H (2016) Analysis of cloud computing attacks and countermeasures. In: Proceedings of the 18th international conference on advanced communication technology (ICACT), Pyeongchang, South Korea
39. Atya AOF, Qian Z, Krishnamurthy SV, Porta TL, McDaniel P, Marvel L (May 2017) Malicious co-residency on the cloud: attacks and defense. In: Proceedings of the IEEE Conference on Computer Communications, Atlanta, pp 1–9
40. Liu B, Yu XL, Chen S, Xu X, Zhu L (2017) Blockchain based data integrity service framework for iot data. In: Proceedings of the IEEE international conference on web services, Honolulu, pp 468–475
41. Original scrypt function for tarsnap. http://www.tarsnap.com/scrypt.html. Accessed 10 April 2018
42. X11 official documentation for dash. https://dashpay.atlassian.net/wiki/spaces/DOC/pages/1146918/X11. Accessed 10 April 2018
43. List of high profile cryptocurrency hacks so far. http://storeofvalueblog.com/posts/cryptocurrency-hacks-so-far-august-24th/

Malware Analysis Using Image Classification Techniques

4

Ajay Singh, Anand Handa, Nitesh Kumar and Sandeep Kumar Shukla

Abstract

These days, almost every device like mobile phones, laptops to large systems such as power grid and nuclear plants are subjected to cyberattacks. Among serious cyber threats, malware-borne threats evolve daily and have the capacity to disrupt both IT and OT systems. A typical antivirus software uses primitive approaches such as generation of signatures of known malware beforehand and then comparing newly downloaded executables against these signatures to detect malware. In recent years, the malware authors have been highly successful in evading signature-based detection techniques. However, machine-learning-based malware detection and classification have gained a lot of importance in recent time. Machine learning methods extract features from binaries using different types of analyses. Static analysis does not execute the binary, but parses the binary to extract features such as use of APIs, size of different sections, etc. The malware authors evade static-analysis-based feature extraction by code obfuscation, packing, and encryption. Therefore, dynamic analysis techniques extract features by letting the code execute in a sandbox and collecting information on runtime activities. The dynamic analysis techniques can be somehow evaded by detecting the sandbox environment and not executing any abnormal or malicious activities inside the sandbox. Therefore, there is an urgent need to find a new approach to overcome the shortcomings of static or dynamic analysis. In this chapter, we discuss an approach to analyze malware for Windows and Linux executables using image representation of the binaries.

Keywords

Malware classification • Convolutional neural network • Machine learning • Deep neural network • Image processing

A. Singh · A. Handa (✉) · N. Kumar · S. K. Shukla
Indian Institute of Technology Kanpur, Kanpur, India
e-mail: ahanda@cse.iitk.ac.in

A. Singh
e-mail: ajay199109@gmail.com

N. Kumar
e-mail: niteshkr@cse.iitk.ac.in

S. K. Shukla
e-mail: sandeeps@cse.iitk.ac.in

4.1 Windows Malware Classification Using Image Representation

The increasing use of the Internet has made our lives convenient but exposed us to the risk of being exploited. Illegitimate users try to implement large-scale malware attacks to exploit the vulnerabilities of the world's most widely used operating system or software ecosystem, i.e., Microsoft Windows. Windows is the most attacked operating system, with over 67% of all malware attacks in 2017 being aimed at systems having the Windows OS. The other operating system gaining popularity in IT and OT systems is Linux. Thus, Linux malware has started to become abundant.

Traditional AV (antivirus) software and malware identification tools use signature-based mechanisms. Though these signature-based mechanisms for detecting malware are effective, these mechanisms fail in case attackers apply obfuscation techniques or when some new malware arrives. VirusTotal [1] witnesses a submission of more than 80,000 files every day for analysis. With such high volume, manual analysis is not possible. So there is a need to automate the analysis process.

The actual motivation for this work came from a quote: "A picture is worth a thousand words," by Fred R. Bernard. In this work, we apply the same and check whether it holds for malware analysis or not. We implement a visualization-based approach for malware classification where an executable is converted into a visual representation, and then we analyze the visible patterns in the image. The executables having a similar structural code are classified into one class. First, we use grayscale image representation but the accuracy is not

S. K. Shukla and M. Agrawal (eds.), *Cyber Security in India*, IITK Directions 4,
https://doi.org/10.1007/978-981-15-1675-7_4

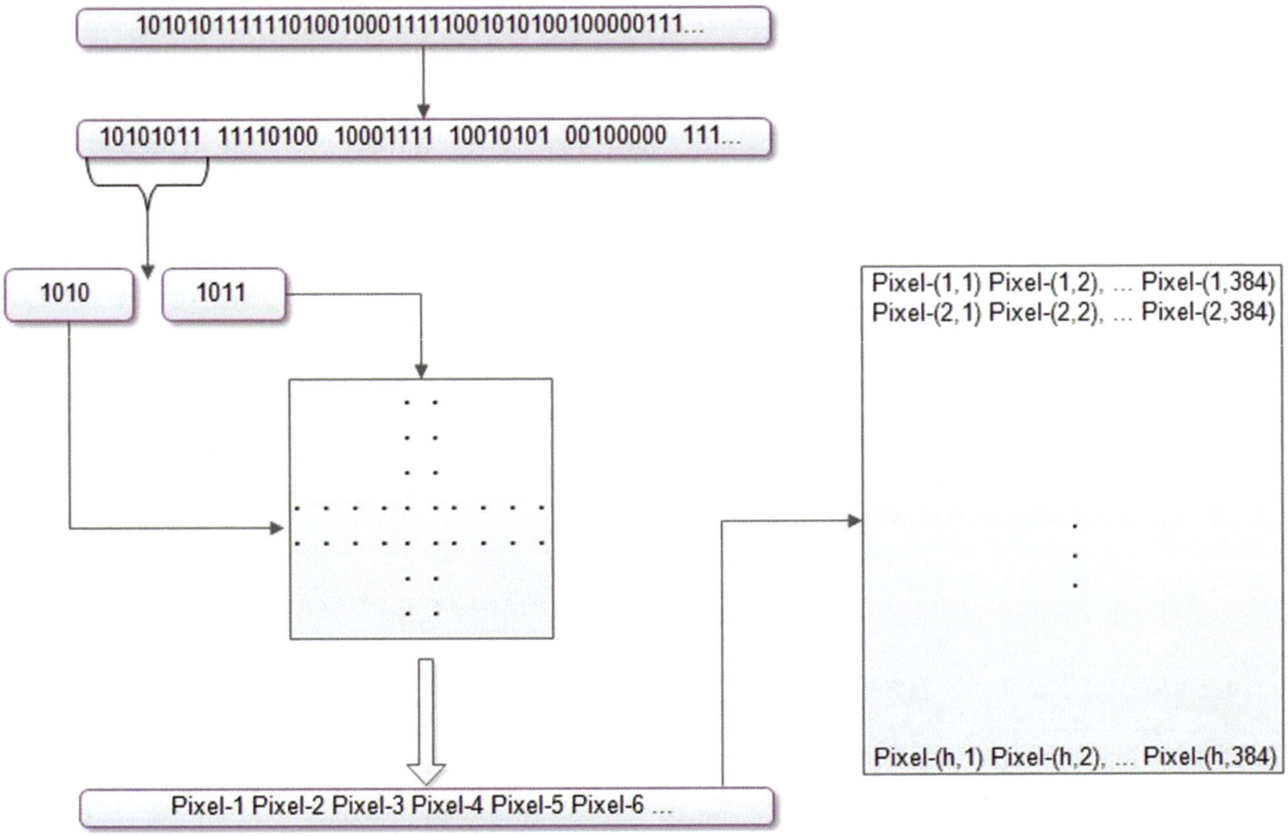

Fig. 4.1 Conversion of Windows and Linux executables to RGB image

very good. Hence, we use RGB representation by which we make a significant improvement in the accuracy which is highlighted in Sect. 4.1.2.

4.1.1 Data Collection and Labeling

We collected more than 60,000 samples from various known malware repositories such as Malshare [2], VirusShare [3], and VirusTotal [1]. We removed the duplicates by comparing the hash values and were left with 37,374 samples belonging to 22 families and 8 classes. Since we used supervised learning, we needed to label the malicious executables before performing the learning, VirusTotal report helped in labeling them by directly querying VirusTotal.

4.1.2 Data Generation

After labeling of samples, we perform the data generation. In data generation, all the executables are considered as a sequence of 0s and 1s and then divided into 8-bit units. We use a two-dimensional color map to store the RGB values corresponding to each byte. This process is repeated for every 8-bit unit, and a sequence of pixel values is generated which is stored in the form of a two-dimensional matrix to get an image visualization. The width of the matrix is kept fixed at 384 bytes and the height is variable as it directly depends on the size of the binary. Figure 4.1 depicts the method of conversion of a Windows executable to RGB image and Fig. 4.2 represents the images of a few Windows malware belonging to different families, namely, Trojan, Virus, and Backdoor.

4.1.3 Classification

For classification, we use a deep CNN and residual neural network. The images are resized into a size of 32×32. The prepared dataset for eight classes is split into 26,149 training and 11,225 testing samples for validation. The deep CNN model and ResNet-50 achieve an accuracy of 98.98% and 99.40%, respectively. To mention here that for implementation of our work, we use a variety of Python libraries such as

Fig. 4.2 Sample images of windows malware belonging to different families

Table 4.1 CNN architecture configuration detail

#	Layer (type)	Output shape	Param#	#	Layer (type)	Output shape	Param#
1	Conv2D	(None, 32, 32, 32)	832	9	Conv2D	(None, 1, 1, 120)	300120
2	MaxPooling	(None, 7, 7, 32)	0	10	MaxPooling	(None, 1, 1, 120)	0
3	Conv2D	(None, 7, 7, 50)	40050	11	Dropout	(None, 1, 1, 120)	0
4	MaxPooling	(None, 2, 2, 50)	0	12	Flatten	(None, 120)	0
5	Conv2D	(None, 2, 2, 80)	200100	13	Dense	(None, 512)	61952
6	MaxPooling	(None, 1, 1, 80)	0	14	Dropout	(None, 512)	0
7	Conv2D	(None, 1, 1, 100)	300120	15	Dense	(None, 22)	11286
8	MaxPooling	(None, 1, 1, 100)	0	16	N/A	N/A	N/A

The total number of parameters are 714,420, Trainable parameters are 714,420
Non-trainable parameters are 0

PIL for image generation, *Keras* and *TensorFlow* for model learning, and *Seaborn* and *matplotlib* for visualization and plotting graphs. Table 4.1 shows the configuration detail of the CNN and Fig. 4.3 represents the ResNet-50 models used for training and testing.

4.1.4 Packed and Unknown Malware Classification

In this work, we also perform experiments on packed malware. We create a dataset of 714 test samples of packed malware using a packer [6] which encrypts the executables and 10,961 test samples of unknown malware. The models (deep CNN and ResNet-50) are then tested on the packed dataset and achieves an accuracy of 60.5% and 53.22%, respectively, and for unknown malware dataset, the reported accuracies are 76.97 and 72.50%.

4.1.5 Results

The corresponding precision, recall, and F1-score for CNN and ResNet-50 on our dataset for windows executables as well as for the packed malware executables and unknown malware are shown in Tables 4.2 and 4.3.

4.2 Linux Malware Classification Using Image Representation

Over the past few years, security threats to Linux devices are on the rise. Mirai botnet attacks turned millions of Linux devices into a zombie army and used them to attack critical infrastructure via Distributed Denial of Service (DDoS). These attacks were an eye-opener for the Linux community. According to AV-Test, though the largest increase in malware targeting in 2016 was MacOS computers, Linux OS computers were close behind with a 300% rise from the previous year. According to another report by WatchGaurd's Internet Security Report, Linux malware threats are among the top 36% of threats identified in Q1 2017. There has been a rapid rise in the submission of executables or different files to antivirus (AV) engines for analysis. Hence, it has become relatively difficult to analyze each file manually. In this work, we extend our image-based approach [7] to classify Linux malware.

4.2.1 Data Collection and Labeling

We collected more than 20,000 samples from various well-known malware repositories such as Malshare, ViruShare, and VirusTotal and then remove the duplicates by comparing the

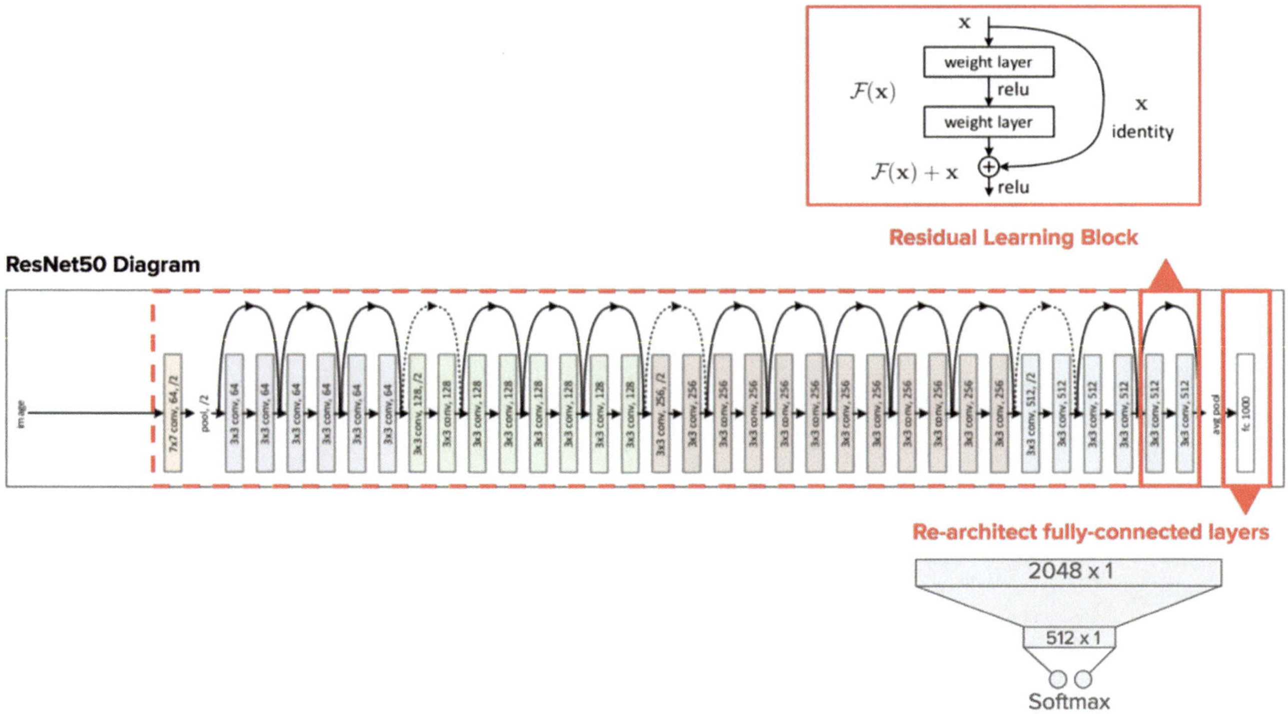

Fig. 4.3 A building block of ResNet-50 and complete architecture. The number of MLP layers might differ. Many such blocks are stacked with varying MLP layers [4,5]

Table 4.2 Experimental results

Classification model	Malware dataset			Accuracy (%)	Malimg dataset			Accuracy (%)
	Samples	Train set	Test set		Samples	Train set	Test set	
CNN	37374	26149	11225	98.98	9339	6527	2812	96.08
ResNet-50	37374	26149	11225	99.40	9339	6527	2812	98.10

hash values and were left with 19,056 samples which belong to five different classes of multiple families. VirusTotal reports were used in labeling them.

4.2.2 Data Generation

The methodology adopted for data generation of ELF executables is similar to the one adopted for a Windows executable as discussed in Sect. 4.1.2. Figure 4.1 depicts the method of conversion of an ELF executable to RGB image and Fig. 4.4 represents the images of Linux malware belonging to different families, namely, Backdoor, Virus, and Botnet.

4.2.3 Classification

For classification, we use a deep CNN and residual neural network as discussed in Sect. 4.1.2. The prepared dataset for five classes is split into 13,338 training and 5,718 testing samples for validation. The deep CNN model and ResNet-50 achieve an accuracy of 97.4% and 98.2%, respectively.

4.2.4 Packed Malware Classification

In this work, we also perform experiments on packed malware. We create a dataset of 1000 test samples of packed malware using a packer which encrypts the executables. The models (deep CNN and ResNet-50) are then tested on the dataset and achieve an accuracy of 58.4% and 55.2%, respectively.

4.2.5 Results

The corresponding precision, recall, and F1-score for CNN and ResNet-50 on our dataset for ELF executables as well as for the packed malware executables are shown in Tables 4.4 and 4.5.

Table 4.3 Experimental results for packed and unknown malware

Classification model	Packed malware Test samples	Accuracy (%)	Unknown malware Test samples	Accuracy (%)
CNN	714	60.50	10961	76.97
ResNet-50	714	53.22	10961	72.50

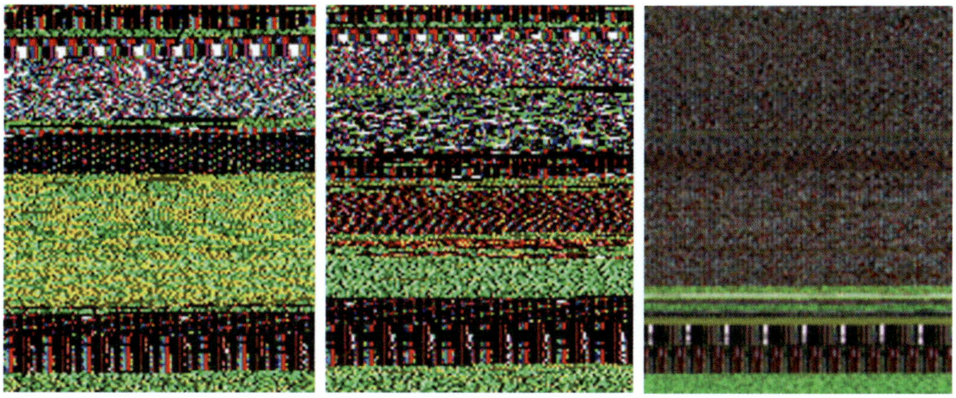

Fig. 4.4 Sample images of Linux malware belonging to different families

Table 4.4 CNN and ResNet results

	CNN results on our dataset			ResNet-50 results on our dataset			
Class	Precision	Recall	F1-score	Class	Precision	Recall	F1-score
Backdoor	0.95	0.99	0.97	Backdoor	0.99	0.98	0.98
Trojan	0.95	0.95	0.95	Trojan	0.98	0.98	0.98
Virus	0.98	0.94	0.96	Virus	0.97	0.97	0.97
DDos	0.99	0.99	0.99	DDos	0.99	0.99	0.99
Botnet	1	0.98	0.99	Botnet	0.98	0.98	0.98
Avg/total	0.97	0.97	0.97	Avg/total	0.98	0.98	0.98

Table 4.5 Experimental results for previously unseen packed malware

	CNN results on our dataset			ResNet-50 results on our dataset			
Class	Precision	Recall	F1-score	Class	Precision	Recall	F1-score
Backdoor	0.48	0.48	0.48	Backdoor	0.46	0.34	0.39
Trojan	0.49	0.81	0.61	Trojan	0.35	0.77	0.48
Virus	0.85	0.78	0.81	Virus	0.82	0.77	0.79
DDos	0.55	0.2	0.29	DDos	0.58	0.15	0.24
Botnet	0.55	0.58	0.56	Botnet	0.55	0.48	0.51
Avg/total	0.58	0.57	0.55	Avg/total	0.55	0.5	0.48

4.3 Conclusion and Future Work

This work shows how a visualization approach works effectively to classify either Linux or Windows malware into their classes. In this work, we use an RGB color map representation rather than using a grayscale representation for our collected Linux and Windows executables. These executables are collected from various known repositories and labeled using VirusTotal reports. After labeling, we convert them into the image representation of their classes by considering the executables as a sequence of 0s and 1s. Lastly, deep CNN and ResNet-50 models are used to train and test our prepared dataset. The models are also tested on packed and previously unseen malware samples, and the accuracies are reported in the results section. This model can be used for classification of malware which is in a hidden format either in PDF or document files or in an encrypted format.

References

1. Lei A, Cruickshank H, Cao Y, Asuquo P, Ogah CPA, Sun Z (2017) Blockchain-based dynamic key management for heterogeneous intelligent transportation systems. IEEE Internet Things J 4(6):1832–1843
2. Malshare- malware repository (2012). http://malshare.com/
3. Han D, Kim H, Jang J (2017) Blockchain based smart door lock system. In: Proceedings of the international conference on information and communication technology convergence (ICTC), Jeju Island, South Korea, pp 1165–1167
4. The full story of the 2018 BSNL India hack (2018). https://medium.com/@kmskrishna/the-full-story-of-the-2018-bsnl-india-hack-85c98e3f10f8
5. Airbnb engineering & data science - image classification (2018). https://medium.com/airbnb-engineering/categorizing-listing-photos-at-airbnb-f9483f3ab7e3
6. Packer-tool upx 3.95 (2018). https://github.com/upx/upx/releases/tag/v3.95
7. Singh A, Handa A, Kumar N, Shukla SK (2019) Malware classification using image representation. In: International symposium on cyber security cryptography and machine learning. Springer, Berlin, pp 75–92

A Review: Malware Analysis Work at IIT Kanpur

5

Amit Kumar, Mugdha Gupta, Gaurav Kumar, Anand Handa, Nitesh Kumar and Sandeep Kumar Shukla

Abstract

The number of newly discovered malware is growing exponentially and pose big threats to the digital world. Users are now more frequent to use the Internet while banking or shopping online. These involve currency transactions and attract malware authors to attack servers and client machines. These hosts hold sensitive information such as personal data, browsing history, shopping history, financial details and much more. A conventional anti-malware software expects malicious programs to contain fixed and known structures. Whereas advanced malware like metamorphic malware is capable of obfuscating their internal structures after each infection. In this chapter, we discuss methods for the detection of such advanced malware using various machine learning techniques which detect/classify malware from static, dynamic and memory forensic features. In Sect. 5.1, we describe our tool *PeerClear* that is designed to detect peer-to-peer botnet. In Sect. 5.2, we describe malware classification tool we designed that works on features extracted from initial stages of their execution. In Sect. 5.3, we describe memory forensics-based malware detection technique we recently developed.

Keywords

Malware classification • Machine learning • Cuckoo sandbox • Bot-net • Memory forensics • volatility tool

5.1 PeerClear: Peer-to-Peer BotNet Detection

According to a report [1], Cybercrimes in India between 2005–2014 rose 19 times. These were those crimes that were actually reported while many of them remain unexposed. In the recent past, bots emerged as one of the most significant players in many cybercrimes. A bot is an infected machine that is under the control of a malicious agent or an attacker and the network consisting of such infected machines is known as botnet. This work presents a new approach for detecting P2P botnet and compares the performance of our method with the other methods in the literature.

A network of infected hosts receiving commands from a command and control (C&C) server is known as Botnet. Earlier botnets used a centralized C&C server architecture, but having single point of failure, led to the advent of another botnet architecture known as peer-to-peer botnets. In this work, an approach known as *PeerClear* for P2P botnet detection is designed. It is a two-step process as shown in Fig. 5.1. The first step comprises host detection and detects those hosts which are involved in P2P traffic. In the second step, the hosts detected in the first stage are subjected to botnet detection. We implement previous approaches and compared with PeerClear. *PeerClear* uses a flow-based approach and it outperforms the other approaches with a high detection rate of 99.85%.

The earlier approaches [2] are not able to categorize P2P traffic, but they are able to classify or detect the P2P botnet. They do not differentiate between P2P and non-P2P traffic. In our work, we categorize the P2P traffic using all the features of a P2P host and a separate P2P bot detection module. The

A. Kumar · M. Gupta · G. Kumar · A. Handa (✉) · N. Kumar · S. K. Shukla
Indian Institute of Technology Kanpur, Kanpur, India
e-mail: ahanda@cse.iitk.ac.in

A. Kumar
e-mail: amitkr@cse.iitk.ac.in

M. Gupta
e-mail: mugdhag@cse.iitk.ac.in

G. Kumar
e-mail: 0202gaurav@gmail.com

N. Kumar
e-mail: niteshkr@cse.iitk.ac.in

S. K. Shukla
e-mail: sandeeps@cse.iitk.ac.in

S. K. Shukla and M. Agrawal (eds.), *Cyber Security in India*, IITK Directions 4,
https://doi.org/10.1007/978-981-15-1675-7_5

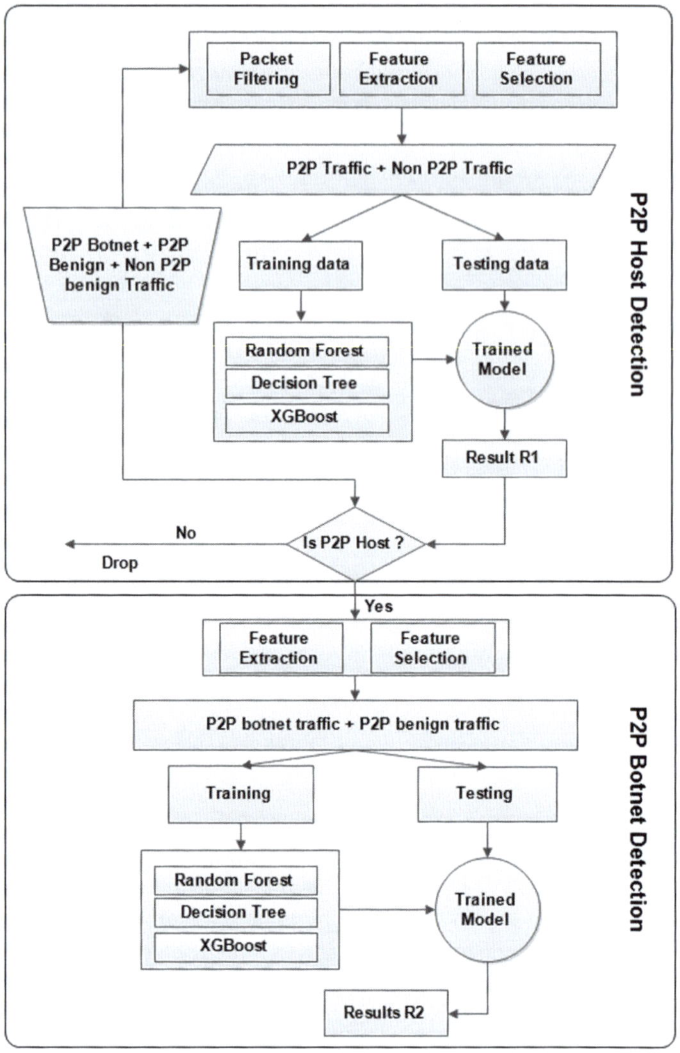

Fig. 5.1 Flow chart of our approach to detect P2P botnet

P2P bot detection module uses P2P botnet salient features for bot detection.

Three types of dataset are collected for performing the experiments. They are - P2P benign network traffic, P2P botnet traffic, and non-P2P traffic. The dataset is in pcap file format, which consists of network traffic information through packet capture. The dataset types, as mentioned earlier, are collected from different sources belonging to distinct hosts and botnets. P2P benign network traffic is collected from 11 hosts. These hosts executed different P2P benign applications like (Skype, eMule, μ-Torrent, Frostwire, and Vuze.) P2P botnet Traffic consists of traffic from Storm [3], Waledac [4], Zeus [5], and Vinchuca botnet [6]. For non-P2P traffic, the departmental traffic is observed for 5 days.

5.1.1 P2P Host Detection

This step consists of four modules-packet filtering, feature extraction, feature selection, and classification. The goal is to detect hosts who are engaged in P2P activity. The packet filtering module filters the packets which are unwanted such as multicast packets, broadcast packets, and DNS generated traffic packets, and the remaining are forwarded to the feature extraction module. The filtering is done to reduce the overhead for packet monitoring and the processing time for each packet. In feature extraction, the prominent features which distinguish the P2P hosts from Non-P2P are extracted from the pcap files using tshark tool [7]. The features are extracted using such properties as—*Failed connections, DNS filter, and Destination diversity*. After feature extraction, we apply dimensionality reduction to reduce the dimensions of the feature vector using Information Gain algorithm. Finally, we select ten features on which we run machine learning classifiers such as XGBoost, Random Forest, and Decision

Tree. The dataset is split into two groups; the first group consists of 70% of data used for training the classifiers, and the second group consists of 30% of data used for testing the classifiers.

5.1.2 P2P BotNet Detection

In this phase, botnet detection is performed over the identified P2P hosts. We have three modules in this phase - feature extraction, feature selection, and classification. In the feature extraction module, a set of eighteen features of two categories-host access features and flow size features are extracted by *pyshark* [8]. In feature selection, we reduce the dimensionality of feature vector. Information gain is applied and top 2 to 18 features with highest scores are selected. After feature selection, we run same set of machine learning classifiers on the selected features with highest information gain scores. Random Forest classifier achieves the highest accuracy of 99.99% on top 6 features. We train the model with traffic from different sets of botnets and tests by using unseen traffic from different botnet.

5.1.3 Experimental Results

In this work, a total of 7,11,149 P2P botnets, and 8,15,659 benign P2P traffic flows are used for the model. For P2P host detection, the results in terms of True Positive Rate (TPR), False Positive Rate (FPR), Precision, and Accuracy are shown in Table 5.1. Similarly, for P2P Botnet detection the results are presented in Table 5.2.

We compare our work with various previous works and also re-implement their methodology and tested on our traffic flows to validate the reported accuracies, shown in Table 5.3. The overall performance of our system is determined by passing the entire traffic into P2P host detection module. The total traffic consists of non-P2P traffic and P2P traffic. The first module separates P2P hosts from non-P2P hosts. The filtered P2P hosts can be P2P botnet hosts or P2P benign hosts. This traffic is now fed into the other module, i.e., P2P botnet detection module. P2P Botnet detection module distinguishes P2P botnets from P2P benign traffic. The overall accuracy of *PeerClear* is 99.85.

5.2 Malware Classification Using Early-Stage Behavioral Analysis

According to a report by AV-Test institute [12], a total of 250,000 malicious samples get registered every day and reverse engineering the samples for analysis is a complicated and inconvenient task for security analysts. Therefore there is an urgent need to automate this analysis and to minimize the human intervention. The signature-based mechanism fails when a newly created malware or a zero-day malware comes into existence and whose signature is unavailable. To overcome these limitations, in this work, we propose a hybrid approach, a combination of static and dynamic analysis.

The proposed system is implemented to classify the malicious executables into its classes by considering the behavioral data for 4 s. We use feature engineering and supervised machine learning algorithms to achieve good classification accuracy. Figure 5.2 shows the architecture of our classification system.

5.2.1 Dataset

We gathered nearly 0.12 million malicious Windows executables from various available malware repositories such as CDAC Mohali [13], Malshare [14] and VirusShare [15]. We crosscheck the reports by submitting them on Virustotal [16] to ensure that the samples are valid malicious files. Virustotal, in its report, produces results by 70 Antivirus engines. We keep only those samples which are identified as malicious by Microsoft AV engine in the virus total report and label them. We drop those executables which Microsoft AV engine failed to detect. Table 5.4 shows the details of our final dataset.

For data generation, we use Cuckoo [17]. It is a tool that performs malware analysis in a controlled environment and generates the reports containing the behavior of the samples. The Cuckoo architecture consists of a host machine which is responsible for managing the analysis and a set of guest machines which execute the samples.

5.2.2 Feature Extraction, Features Selection, and Classification

We extract features from static analysis and dynamic analysis. Python's pefile library is used to extract features for static analysis of Windows executable. The extracted features are *File header information:* `Machine`, `number_of_sections`, `compile_date`, `number_of_import_sysmbols`, *Optional header information*: `major_version`, `minor_version`, `debug_size`, `check_sum`, and `iat_rva`. Other information are section entropy, size of raw data, size of uninitialized data, section virtual address, and total size of PE file. For dynamic analysis feature extraction, we use Cuckoo sandbox. Features are extracted from the Cuckoo report after the file execution. We extract network-related information, API bins, process-related information, and the signatures provided by Cuckoo sandbox as features. A total of 52 and 78 features are extracted during static and dynamic analysis respectively.

Out of these, all features may not be important for training the machine learning classifiers. Hence, we use information gain algorithm for feature selection. In static analysis, top 10–50 features are selected according to their info-gain score. Similarly, in dynamic analysis, top 10–78 features are selected. Finally, we are left with top 50 (from static analysis) and top 40 (from dynamic analysis) features as a final feature vector.

For classification, the dataset is split into 70–30% for training and testing our model. Different machine learning classifiers like Random Forest, Decision Tree, XGBoost, Simple Neural Network, and k-NN are used to perform classification. For parameter tuning, we apply tenfold cross-validation.

5.2.3 Experimental Results

For static analysis, Random Forest classifier achieves the highest accuracy, which is 97.95% with a False Positive Rate (FPR) of 0.5% as shown in Table 5.5. For dynamic analysis, we train and test our model on the same samples and achieve the highest accuracy of 99.13% using Random Forest classifier with FPR 0.2%. Table 5.6 shows the corresponding results.

There exist a few limitations for dynamic approach as well as for the static approach. Hence, we combine both the approaches to use a hybrid approach. In the hybrid analysis, selected static and dynamic features are combined to train and test the model. Random Forest classifier achieves the highest accuracy of 99.74% with FPR 0.1% for the hybrid model, and the results are presented in Table 5.7.

We execute a total of 1500 samples for 4-s, 8-s, and 12-s, respectively. The same features set is used to train and test the

Table 5.1 P2P host detection results

Classifier	TPR (%)	FPR (%)	Precision (%)	Accuracy (%)
Random forest	99.91	0.003	99.98	99.93
Decision tree	99.89	0.001	99.92	99.88
XGBoost	99.73	0.001	99.92	99.78

Table 5.2 P2P botnet detection result

Classifiers	TPR (%)	FPR (%)	Precision (%)	Accuracy (%)
Random forest	99.98	0.002	99.99	99.99
Decision tree	99.97	0.004	99.97	99.97
XGBoost	99.77	0.024	99.97	99.88

Table 5.3 Accuracy (%) comparison

Authors	Dataset	Approach	Author reported accuracy (%)	Authors proposed model accuracy on our traffic
Pratik Narang et al. [2]	P2P botnet conversations - 50000 P2P benign conversations - 50000	Conversation based	Approx 97%	98.37%
Hojjat et al. [9]	P2P botnet flows - 3296 P2P benign flows - 1233	Flow based	99.26%	99.77%
Himanshi et al. [10]	P2P botnet packets - 41941536 P2P benign packets - 25913400	Flow based	Approx 99%	99.72%
Alauthaman et al. [11]	P2P botnet control packets - 114087 P2P benign control packets - 331526	Flow based	99.20%	94.25%
Our approach	P2P botnet flows - 711149 P2P benign flows - 815659	Flow based	99.85%	N/A

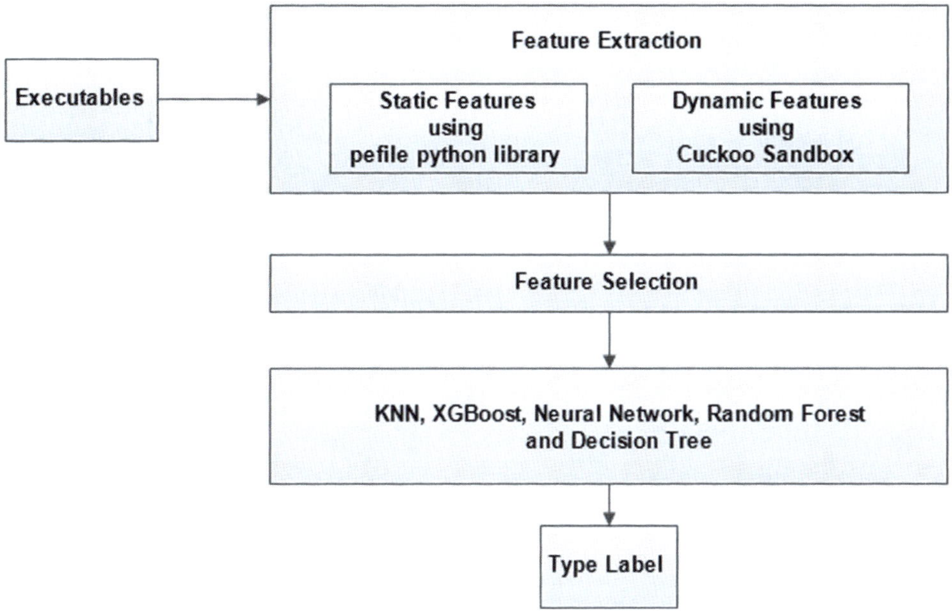

Fig. 5.2 Architecture of our classification system

Table 5.4 Dataset

Malware class	Malware family	Number of samples
Backdoor	Rbot	825
	Win32_Agent	843
	Win32_Small	436
Trojan	Comame!gmb	873
	startpage	1007
	Win32_Bulta!rfn	625
TrojanDownloader	Renos	811
	Small	885
	Win32_Tugspay.A	1052
TrojanDropper	Sventore.C	1154
	Win32_Sventore.A	1203
Virus	Sality	298
	Win32_Krepper.30760	1127
	Win32_Luder.B	1088
Worm	Win32_Allaple.A	1056
	Win32_VB.AT	948
	Win32_Yuner.A	906
Total		15137

various machine learning classifiers. We achieve the highest accuracy of 98.55% on top-40 features for 4-s time interval analysis. For 8-s time interval analysis, we achieve an accuracy of 98.61% on top-50 features, and 12-s time interval analysis, we get an accuracy of 98.65% on top-70 features. From the results, we derive though there is a minor improvement in the accuracy, there is a significant difference in the time interval and the number of features. Hence, we select 4-s as a time interval for performing behavioral analysis.

We also test the performance of our model to classify obfuscated and packed malware samples. To pack the samples, we use a packer [18] and for obfuscation of malware samples, we use metame [19] which is a metamorphic code engine. Random Forest classifier gives the highest accuracy of 96.31% for obfuscated and 96.73% for packed malware using hybrid analysis. Tables 5.8 and 5.9 show the corresponding results.

We also compare our work with other author's work who classify malware into its families/types and also with those

Table 5.5 Experimental results for static analysis

No. of features	Random forest	XGBoost	Decision tree	k-NN	Neural network
10	96.807	96.015	95.795	94.892	90.795
20	97.049	96.323	96.125	**95.046**	93.592
30	97.225	96.367	96.169	94.87	94.869
40	97.827	96.609	95.905	94.738	**96.322**
50	**97.952**	**96.786**	**96.213**	94.738	96.014

Table 5.6 Experimental results for dynamic analysis

No. of features	Random forest	XGBoost	Decision tree	k-NN	Neural network
10	96.585	95.971	96.169	94.254	94.054
20	98.983	98.283	97.952	**95.839**	96.939
30	99.121	98.943	98.063	90.225	98.547
40	**99.135**	99.075	98.437	90.819	**98.855**
50	99.102	99.031	98.085	90.401	98.569
60	99.069	99.119	98.217	90.291	98.414
70	99.069	99.075	**98.481**	90.291	98.723
78	99.04	**99.119**	98.459	90.291	98.679

Table 5.7 Experimental results for hybrid analysis

No. of features	Random forest	XGBoost	Decision tree	k-NN	Neural network
90	**99.736**	99.634	99.185	93.813	99.273

Table 5.8 Accuracy % for obfuscated binaries using various classifiers and approaches

Static analysis				
Random forest	XGBoost	Decision tree	k-NN	Neural network
90.62	88.33	87.78	71.31	88.17
Dynamic analysis				
Random forest	XGBoost	Decision tree	k-NN	Neural network
93.17	92.04	90.62	84.94	91.56
Hybrid analysis				
Random f orest	XGBoost	Decision tree	k-NN	Neural network
96.31	95.17	91.48	85.22	92.79

Table 5.9 Accuracy % for packed binaries using various classifiers and approaches

Static analysis				
Random forest	XGBoost	Decision tree	k-NN	Neural network
92.48	91.79	85.94	78.75	84.36
Dynamic analysis				
Random forest	XGBoost	Decision tree	k-NN	Neural network
92.81	92.48	90.52	84.64	88.15
Hybrid analysis				
Random forest	XGBoost	Decision tree	k-NN	Neural network
96.73	96.40	89.86	86.89	90.12

who attempt to predict malicious behavior with a short duration behavioral data. Table 5.10 shows the comparison.

5.3 Automated Malware Detection Using Memory Forensics

A new type of malware threat is fileless malware. Such malware remains memory resident, never saving the binary of the disk. Such malware detection can not be done by analysis of binaries directly. Memory forensics from memory snapshots of the system may reveal such malware. In dynamic analysis, analysts cannot discover the hidden processes, but these hidden artifacts can be discovered by doing a full memory inspection in memory forensics. Memory forensics is also used to detect the code injection in usual processes and non-persistent malware. We take memory snapshots and examine the memory dump file. In the past few years, we have seen several efforts to automate memory forensics-based malware analysis using machine learning techniques. In this work, we present an improvement on past attempts at automated malware detection using memory forensics.

In this section, we present memory forensics of Windows binaries. We collect malware samples from various repositories such as VirusShare [27], MalShare [14], and VXHeaven [28] and collect benign files from a freshly installed Window 7 machines and Softonic [29], and Softpedia [30]. We use 1159 malware and 1051 benign files for further analysis. For memory dump generation, we use Cuckoo sandbox, which is a tool for performing dynamic malware analysis. We use Ubuntu 18.04 with 64 GB RAM, 1 TB SSD, 6 TB external storage, and INTEL I7 octa-core processor for data generation and feature extraction. It is a challenge to utilize the space because, for 512 samples, we need 6 TB disk space. And for 2210 samples it requires approximately 25 TB. Hence, we modify our script in such a way that it selects one memory dump out of the five generated dumps and delete all other memory dumps. Due to space and time limitations, we perform this analysis in stages. Initially, we generate the dumps and corresponding reports for 5-s and perform the analysis. For 10-s and 15-s time intervals, the same process is repeated. Our methodology is described in four parts: Generation of memory dumps, selection of the most informative memory dump, feature extraction, and classification.

5.3.1 Memory Dump Generation and Selection

For memory dump generation, we use the Cuckoo sandbox. It uses VirtualBox Python API (vboxapi) to save a memory dump in the host machine. Five memory dumps are taken at an interval of 5-s, 10-s, and 15-s, and later one memory dump is selected for further processing. It is a very cumbersome task to select the one memory dump that contains maximum malicious artifacts. The process of choosing one memory dump out of the five generated dumps is based on the number of processes in execution at the time of memory dump generation. Few samples get entirely executed after two or three memory dumps are generated. We drop those memory dumps which are generated after the complete execution of the sample. After the selection of one memory dump, we start feature extraction with the help of the Volatility tool.

5.3.2 Feature Extraction, Selection, and Classification

We use various Volatility tool plugins to extract features from the selected memory dump. A total of 259 features are extracted. Examples of extracted features are *hidden or injected code, hidden or injected dlls, TCP listeners, UDP listeners, inactive processes, hidden processes, unlinked processes, session id, process id, number of handles, number of threads, terminated process, processes having debug privilege, suspicious unknown call backs*, etc. We apply a feature reduction technique for dimensionality reduction based on information gain measure. For 5-s time interval memory dump analysis, we select 45 features, similarly for 10-s analysis 61 features are selected, and for 15-s dump analysis 69 features are selected according to their information gain score. In feature selection, we drop those features whose score is zero. For classification, we use supervised machine learning classifiers such as XGBoost, k-NN (K-Nearest Neighbor), Random Forest, and Decision Tree with the help of Python library Scikit-Learn. To train and test the models, we use a 70 and 30% split of the data.

5.3.3 Experimental Results

The experiments are performed in three stages. In the first stage, we generate dumps for 5-s and select 45 features and then apply a set of machine learning classifiers on top-10 to top-45 features. XGBoost classifier achieves the highest accuracy of 95.84% for top-45 features as shown in Table 5.11. In the second stage, a memory dump for the 10-s interval is generated, and 61 features are selected from which top-10 to top-61 features are used to train and test the same set of machine learning classifiers. Random Forest classifier achieves the highest accuracy of 98.67% for top-30 features. Table 5.12 shows the corresponding results.

In the third stage, 15-s time interval dumps are generated, and 69 features are selected. Random Forest classifier achieves The highest accuracy of 97.89% for top-30 features as shown in Table 5.13. From Tables 5.11, 5.12 and 5.13, it is clear that the accuracy for 10-s memory dump analysis is

Table 5.10 Comparison with existing approaches

Author	Approach	Analysis time	Dataset	Accuracy
[20]	Dynamic	–	3768 samples (13 families)	0.945 ROC area
[21]	Dynamic	5 min	81 malwares 69 benign log files	0.96 ROC area
[22]	Dynamic	4 s 20 s	2345 benign 2286 malicious	91% 96%
[23]	static		745 malwares (6 families) 40 benign	98.0%
[24]	Static	–	350,016 malicious 81,910 benign	95.20%
[25]	Hybrid	–	2939 malicious 541 benign	97.055%
[26]	Hybrid	–	1000 malicious 1000 benign	96.60%
Ours	Static	0.31 s	15137 samples	97.95%
	Dynamic	8.29 s	17 families	99.13%
	Hybrid	8.49 s	6 types	99.74%

Table 5.11 Results for 5-s dump analysis

No. of features	Random forest	XGBoost	Decision tree	KNN
10	95.13	94.23	93.33	86.13
20	95.14	94.68	93.33	86.12
30	94.09	95.28	94.68	86.58
40	93.78	95.29	95.58	86.58
45	95.13	**95.84**	95.13	88.23

Table 5.12 Results for 10-s dump analysis

No. of features	Random forest	XGBoost	Decision tree	KNN
10	96.51	94.65	93.72	93.26
20	96.97	95.12	93.26	94.19
30	**98.67**	94.65	93.26	94.19
40	97.05	95.58	93.72	93.26
50	96.51	95.12	93.72	93.25
61	96.98	95.11	95.58	93.25

Table 5.13 Results for 15-s dump analysis

No. of features	Random forest	XGBoost	Decision tree	KNN
10	95.79	96.29	92.23	90.21
20	96.81	96.22	92.74	90.22
30	**97.89**	96.22	92.23	90.25
40	96.22	96.22	92.74	88.68
50	96.80	94.69	93.84	88.68
60	96.36	95.20	95.86	88.68
69	96.22	95.20	94.85	88.68

Table 5.14 Time analysis results for different memory dumps

Models	5 Dumps generation time (s using Cuckoo	Dump selection time (s)	Feature extraction and prediction using ML time (s	Total time (s)
5-s	60	10	89	159
10-s	84	11	102	197
15-s	110	13	132	255

Table 5.15 Comparison with others approach

Authors	Accuracy (%)	No. of samples	Features used
Mosli et al. (2016)	96	400 malware 100 benign	Registry keys, DLLs, API
Rathnayaka et al. (2017)	90	200 malware 200 benign	Kernel memory, kernel objects, registry, Api, strings, File systems
Masoume et al. (2014)	98	350 malware 200 benign	Registry changes, function calls, etc.
Ours	98.67	1159 malware 1051 benign	Registry, suspicious DLLs, process dump, kernel dump, code injection, File System etc.

highest as compared to 5-s and 15-s memory dump analysis. It may be possible due to the amount of information in 5-s memory dump is less as compared to 10-s and 15-s memory dumps. The features we get in 10-s memory dumps are more discriminating as compared to 15-s memory dumps. We also perform the time analysis for different stages and the corresponding results are shown in Table 5.14. Table 5.15 compares our work against past work based on memory forensics.

5.4 Conclusion

In this chapter, we presented three categories of works related to malware analysis, which are associated with P2P botnet detection, malware classification, and malware detection using memory forensics. Our P2P botnet detection framework achieves the highest accuracy of 99.85% with a low false positive rate. Our malware classification model is based on early-stage behavioral analysis. It classifies malware into six different classes, which are Backdoor, Trojan, TrojanDownloader, TrojanDropper, Worms, and Virus using hybrid analysis. The model achieves the highest classification accuracy of 99.74% for 4-s of behavioral data. Our malware detection model is trained using features extracted from the memory dumps, which are generated by executing the samples in a controlled environment. The model achieves the highest detection accuracy of 98.67% for 10-s time interval memory dump analysis.

References

1. BGR News Report (2016). https://www.bgr.in/news/cyber-crimes-in-india-rose-19-times-between-2005-2014/
2. Narang P, Ray S, Hota C (2014) Peershark: detecting peer-to-peer botnets by tracking conversations. In: IEEE security and privacy workshops
3. Holz T, Steiner M, Dahl F, Biersack E, Freiling F (2008) Measurements and mitigation of peer-to-peer-based botnets: a case study on storm worm. In: Proceedings of the 1st USENIX workshop on large-scale exploits and emergent threats
4. Nunnery C, Sinclair G, Kang BB (2010) Tumbling down the rabbit hole: exploring the idiosyncrasies of botmaster systems in a multi-tier botnet infrastructure. In: Proceedings of the 3rd USENIX conference on Large-scale exploits and emergent threats: botnets, spyware, worms, and more
5. Lelli A (2018) Zeusbot/Spyeye P2P updated, fortifying the botnet. https://www.symantec.com/connect/blogs/zeusbotspyeye-p2p-updated-fortifying-botnet
6. Lontivero: a resilient peer-to-peer botnet agent in .NET. https://github.com/lontivero/vinchuca
7. Tshark - dump and analyze network traffic (2018). https://www.wireshark.org/docs/man-pages/tshark.html
8. KimiNewt: python wrapper for tshark, allowing python packet parsing using wireshark dissectors (2018). https://github.com/KimiNewt/pyshark
9. Beiknejad H, Vahdat-Nejad H, Moodi H (2018) P2P botnet detection based on traffic behavior analysis and classification. Int. J. Comput. Inf. Technol. 6(1):01–12
10. Dhayal H, Kumar J (2017) Peer-to-peer botnet detection based on bot behaviour. Int J Adv Res Comput Sci 8(3)
11. Alauthaman M, Aslam N, Zhang L, Alasem R, Hossain MA (2018) A P2P botnet detection scheme based on decision tree and adaptive multilayer neural networks. Neural Comput Appl 29(11):991–1004
12. Av-TEST security institute (2018). https://www.av-test.org/en/statistics/malware/
13. CDAC Mohali (2018). https://cdac.in/index.aspx?id=mohali
14. Malshare (2018). https://malshare.com/ (2018)

15. Virusshare (2018) https://virusshare.com/
16. VirusTotal (2012). https://www.virustotal.com. Acquired by Google Inc
17. Cuckoo Sandbox (2018). https://cuckoosandbox.org/
18. Packer-tool upx 3.95 (2018). https://github.com/upx/upx/releases/tag/v3.95
19. Metamorphic code engine (2019). https://github.com/a0rtega/metame
20. Nari S, Ghorbani AA (2013) Automated malware classification based on network behavior. In: International conference on computing, networking and communications (ICNC)
21. Tobiyama S, Yamaguchi Y, Shimada H, Ikuse T, Yagi T (2016) Malware detection with deep neural network using process behavior. In: 40th annual IEEE conference on computer software and applications conference (COMPSAC)
22. Rhode M, Burnap P, Jones K (2017) Early stage malware prediction using recurrent neural networks. CoRR arXiv:abs/1708.03513
23. Damodaran A, Troia FD, Visaggio CA, Austin TH, Stamp M (2017) A comparison of static, dynamic, and hybrid analysis for malware detection. J Comput Virol Hacking Tech 13:1
24. Saxea J (2015) Berlin: deep neural network based malware detection using two dimensional binary program features. In: 10th international conference on malicious and unwanted software (MALWARE) (2015)
25. Islam R, Tian R, Batten LM, Versteeg S (2013) Classification of malware based on integrated static and dynamic features. J Netw Comput Appl 36(2):646–656
26. Santos I, Devesa J, Brezo F, Nieves J, Bringas PG (2013) Opem: a static-dynamic approach for machine-learning-based malware detection. In: International joint conference CISIS'12-ICEUTE 12-SOCO 12 special sessions. Springer, pp 271–280
27. Virusshare - malware repository (2011). https://virusshare.com/
28. Vx heaven dataset (2016). https://archive.org/download/vxheaven-windows-virus-collection
29. Softonic (2019). https://en.softonic.com/windows/
30. Softpedia (2019). https://win.softpedia.com/

Honeypot Deployment Experience at IIT Kanpur

6

Rohit Sehgal, Nishit Majithia, Shubham Singh, Sanjay Sharma, Subhasis Mukhopadhyay, Anand Handa and Sandeep Kumar Shukla

Abstract

Honeypot is a entrapment mechanism that provides attackers with all necessary resources needed for a successful attack. Unlike Intrusion Detection System (IDS) and Intrusion Prevention System (IPS), which tries to stop malicious activity after detection, honeypot allows attacker to perform the attack and allows a cybersecurity researcher to get an thorough understanding of the attack patterns. Honeypot also helps in capturing the attack signatures, scripts and the payloads from the system. This chapter provides an idea about honeypots and different types of honeypots. We also describe the analysis of the attacks on honeypots deployed at IIT Kanpur.

Keywords

IoT honeypot • HoneySMB • HoneyFTP • HoneyFARM • MQTT

6.1 Introduction

The growth in the number of the Internet users and in e-commerce fascinated cybercriminals, resulting in, increasing cyberattacks over the past decade. Over 90,000 websites are hacked every day [1]. WordPress is the most hacked CMS with 83% of hacked websites using the WordPress platform. Today, hundreds of more threatening attacks are aimed at the Internet users each day. As per positive technologies Q1 2017 report [2], SQL Injection and Cross-Site Scripting were the most common attacks, each representing almost one-third of the total number of attacks in Fig. 6.1. In 2019 these figures have not changed much.

During the last two decades, many techniques have been deployed to detect malicious activities and preventing legitimate systems like Firewall, Intrusion Detection System (IDS) and Intrusion Prevention System (IPS) and Honeypots. Firewall, IDS and IPS are traditional methods to prevent attacks on legitimate systems. A firewall is a network security system used to monitor incoming and outgoing traffic based on predefined rules, whereas an IDS and IPS are devices or software applications that monitor a network or systems for malicious activity or policy violations. It is possible for an attacker to bypass the firewall if it is not correctly configured. Signature-based IDS unable to detect the zero-day attacks. A honeypot is an entrapment mechanism like a spider web, that attracts attacker to enter, and execute malicious activities. A honeypot is intentionally made vulnerable to attacks to make it easy for an attacker to enter and display his/her hacking skills. The honeypots are usually built with strong isolation from the rest of the IT system so the attacks cannot escape the honeypot into the main network. After the attackers download malicious payload, try various tricks, we can collect forensics to understand the attackers' modus operandi.

In this chapter, we discuss various types of honeypots. They provide us a mechanism to collect the information about attack vectors that identify the collaborators and enables us to answer questions such as how to defend against and defeat the attacker when attacker's identity and prior knowledge of her/his attack strategies are not available [3]. The main advantage of employing honeypot is that it allows us to analyse how the attackers act in a vulnerable system and what techniques they use for exploiting system's vulnerabilities. This analysis provides valuable information to security researchers or network administrator about the skills of the cybercriminals. Figure 6.2 shows an example scenario of honeypot system.

R. Sehgal · N. Majithia · S. Singh · S. Sharma (✉) · S. Mukhopadhyay · A. Handa · S. K. Shukla
Indian Institute of Technology Kanpur, Kanpur, India
e-mail: sanjaysr@cse.iitk.ac.in

S. Mukhopadhyay
e-mail: subhasism@cse.iitk.ac.in

A. Handa
e-mail: ahanda@cse.iitk.ac.in

S. K. Shukla
e-mail: sandeeps@cse.iitk.ac.in

S. K. Shukla and M. Agrawal (eds.), *Cyber Security in India*, IITK Directions 4,
https://doi.org/10.1007/978-981-15-1675-7_6

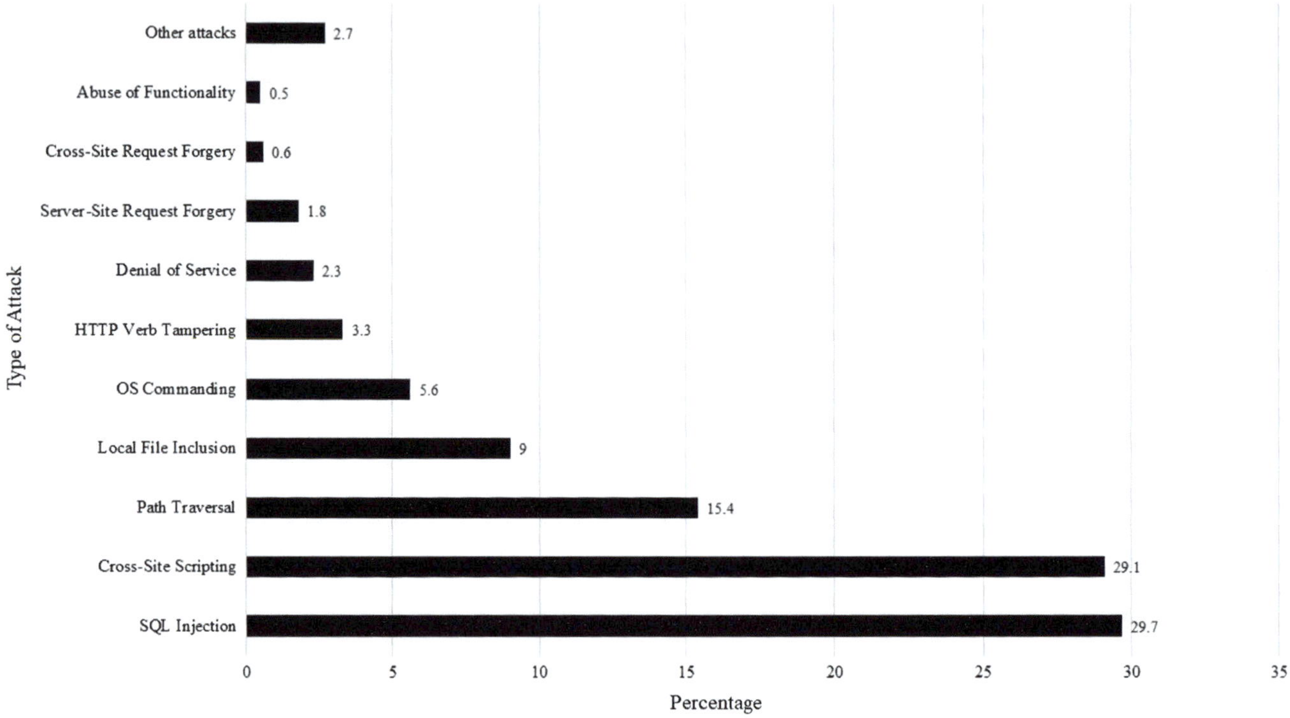

Fig. 6.1 Top 10 attacks on web applications in Q1 2017

Fig. 6.2 An example scenario of honeypot

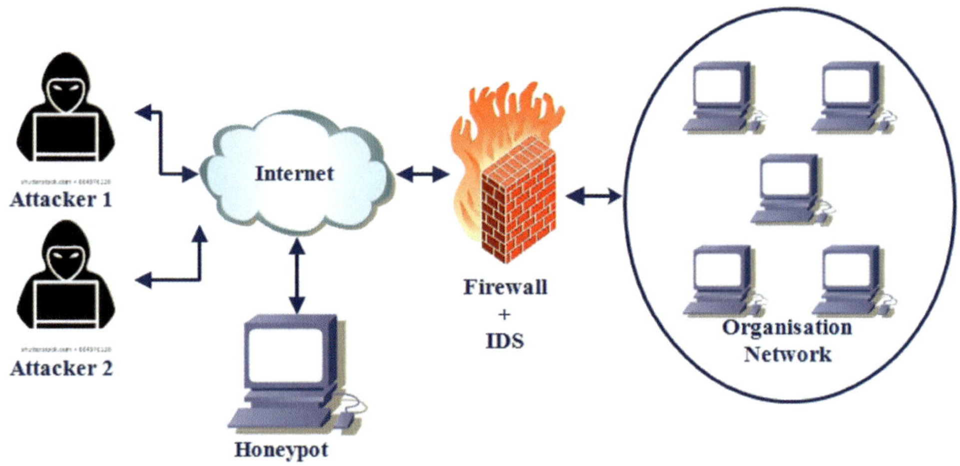

6.2 Classification of Honeypots

Honeypots are further classified along three different dimensions. The first dimension is based on the level of interaction it offers to the attacker, i.e. high and low [4,5]. Another type is based on services provided by it, i.e. the client-side or the server-side. A third classification is based on the utilisation of honeypots, i.e. either these are used for research or production purpose.

6.2.1 Low/High-Interaction Honeypot

Low-interaction honeypots offer frequently used services to an attacker. These are not actual services but emulated, for example, a login screen with two fields username and password. This page is capable of capturing information, but the actual page is missing too which the attacker tries to log in. These types of honeypots are not much effective from the research point of view because of limited interaction.

High-interaction Honeypots provide maximum information about an attacker. However, these are prone to risks if not deployed correctly. This kind of honeypots provides much more information about the attackers. For example, tools used for the attack, attack pattern and payload used. These honeypots are time consuming to build and maintain, but if deployed properly they provide plethora of information and become a valuable asset for the organisation.

6.2.2 Server/Client Honeypot

Server-side honeypots are traditional honeypots to analyse attacks on servers and services provided by them. They provide a mechanism to learn server-side attacks and thwart them in the future. Server side honeypots offer software as well as platform-based vulnerabilities to the attackers and passively wait to be attacked. Server honeypots provide services which are offered by actual servers for example *telnetd* is a type of service provided by server. Server-side honeypots are unable to detect client-side attacks.

For client-side attack detection, client system needs to interact with servers and to process malicious response data. Client-side honeypots are able to analyse attacks initiated by malicious web servers on clients that connect to those servers.

6.2.3 Production/Research Honeypot

Production honeypots are used by companies and organisations to mitigate risks from cybercriminals [6]. These honeypots are usually low-interaction honeypots because of their ease in installation. These are deployed in the organisation network to alleviate actual risks posed to servers. Because of low interaction, these are not too effective, and also from the research point of view these do not provide good amount of information about the attacks.

Research honeypots are high-interaction honeypots, and these are mostly deployed for researching behaviours of attackers. By using these honeypots, a security researcher can get more information about attacks, vulnerabilities and attack patterns used by the attackers.

6.3 Deployed Honeypots

At the Indian Institute of Technology, Kanpur (IITK), we deploy many honeypots for research purpose and extract information from the logs generated by these honeypots and analyse behaviours of the attackers, types of attack and collect payloads for our malware analysis research.

6.3.1 OpenCanary

OpenCanary is a daemon that runs canary services, which trigger alerts when abused. The alerts are sent to a variety of sources, including system log, emails and a companion daemon OpenCanary correlator [7]. Many services offered by OpenCanary are `SSH, FTP, HTTP, HTTP-proxy, MySQL, TELNET, SNMP`, etc.

We deploy OpenCanary honeypot on Ubuntu 18.04 LTS and use its fake services such as **FTP**, **Port Scan**, **SSH**, **HTTP** and **HTTP-proxy**. We analyse the log file generated by OpenCanary honeypot. There are many fields in the log which contain valuable information such as follows:


```
{"dst_host": "10.0.2.4", "dst_port": 22,
"local_time": "2019-05-24
05:55:12.510752", "logdata":
{"LOCALVERSION": "SSH-2.0-OpenSSH_5.1p1
Debian-4", "PASSWORD": "malwarelab",
"REMOTEVERSION": "SSH-2.0-Ruby/
Net:SSH_5.1.0 x86_64-linux-gnu",
"USERNAME": "root"}, "logtype": 4002,
"node_id": "opencanary-1", "src_host":
"10.0.2.15", "src_port": 39463}
```


Some relevant information which we process for further analysis is mention in Table 6.1. We extract information from

Table 6.1 Some relevant information extracted from OpenCanary log

Field name	Description	Value
dst_host	Destination IP	10.0.2.4
dst_port	Destination port no.	22
local_time	Date and time of attack	2019-05-24 05:55:12.510752
LOCALVERSION	Attacker machine service information	SSH-2.0-OpenSSH_5.1p1 Debian-4
PASSWORD	Password tried during SSH attack	malwarelab
REMOTEVERSION	Victim machine or honeypot Machine information	SSH-2.0-Ruby/Net::SSH_5.1.0 x86_64-linux-gnu
USERNAME	Username of attacker machine	Root
src_host	IP of honeypot machine	10.0.2.15
src_port	Port number exploited by attacker	39463

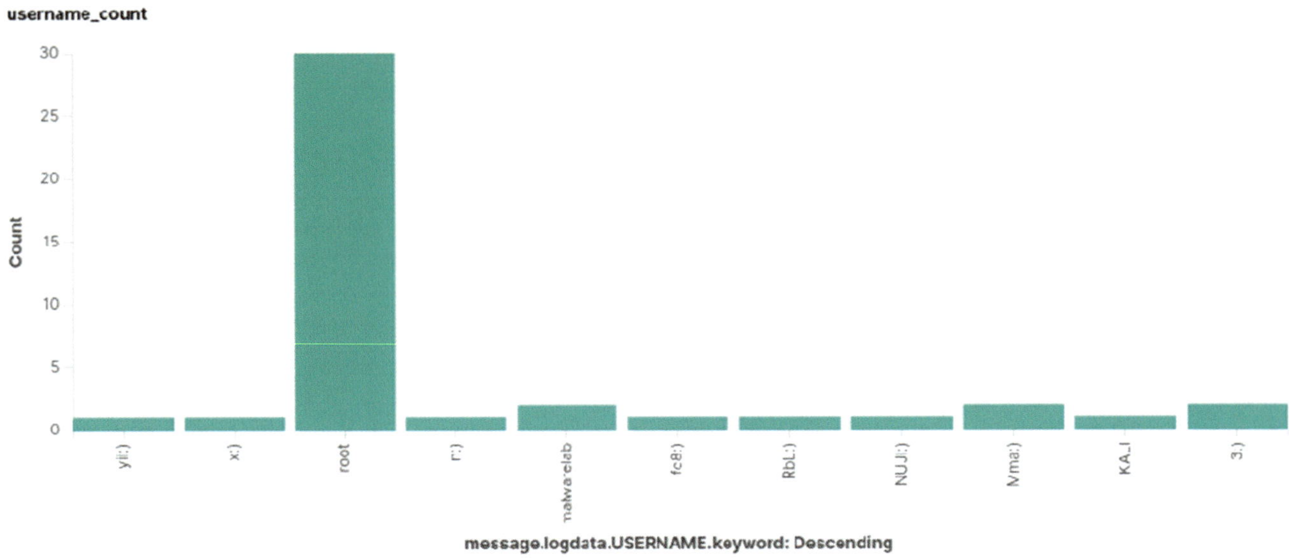

Fig. 6.3 Number of attacks from users

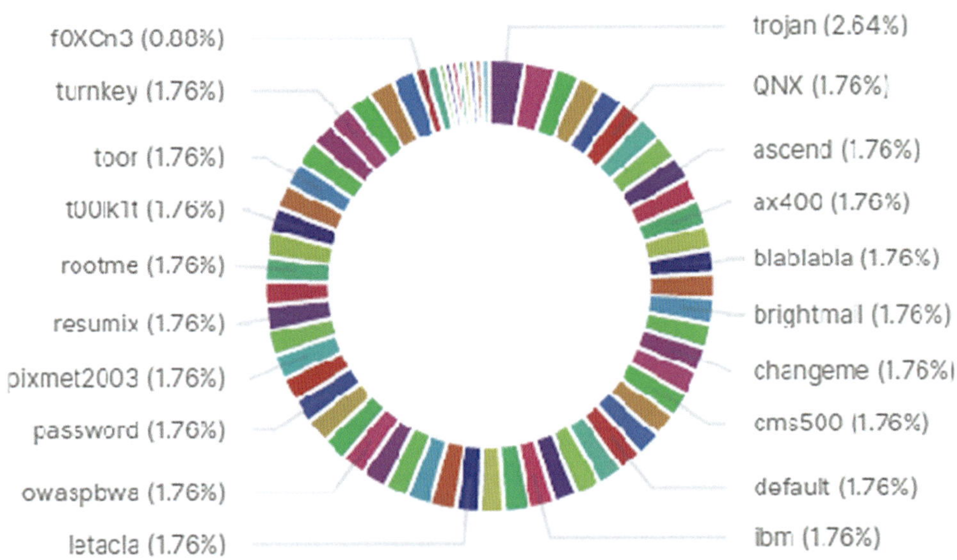

Fig. 6.4 Password used on victim machines

log generated for each service that are enabled in OpenCanary configuration file and analyse this information for further processing. Figures 6.3, 6.4, 6.5 and 6.6 visualise the relevant information using a search and analytics engine i.e. ELK (Elasticsearch, Logstash, and Kibana) [8].

6.3.2 Cowrie

Michel Oosterhof developed Cowrie, a medium-interaction SSH and Telnet Honeypot to log the shell interactions and brute-force attacks by attackers [9]. Some interesting features of Cowrie Honeypot are

- A fake filesystem with the ability to add/remove files.
- Possibility of adding fake file contents so the attacker can cat files such as /etc/passwd. Only minimal file contents are included.
- Logs are stored in JSON format for easy processing.
- Cowrie saves the files downloaded with wget/curl or uploaded with SFTP and SCP for later inspection of logs.

We deploy Cowrie on Ubuntu 18.04 and use SSH service. We analyse the log file generated by Cowrie honeypot. The log file contains the information about activity done by attackers such as `username`, `password`, `message`, `timestamp`, `source IP`, `destination IP`, `source port`, `destination port` and `session detail`. This information is used

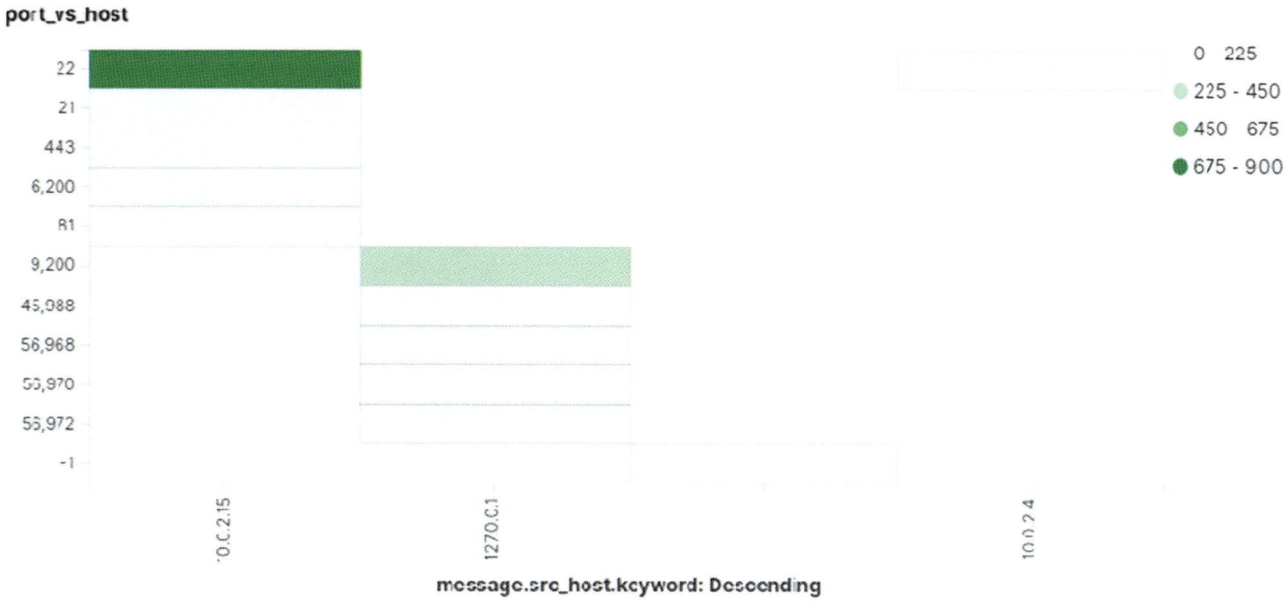

Fig. 6.5 Count of port number accessed by specific IP (attacker IP)

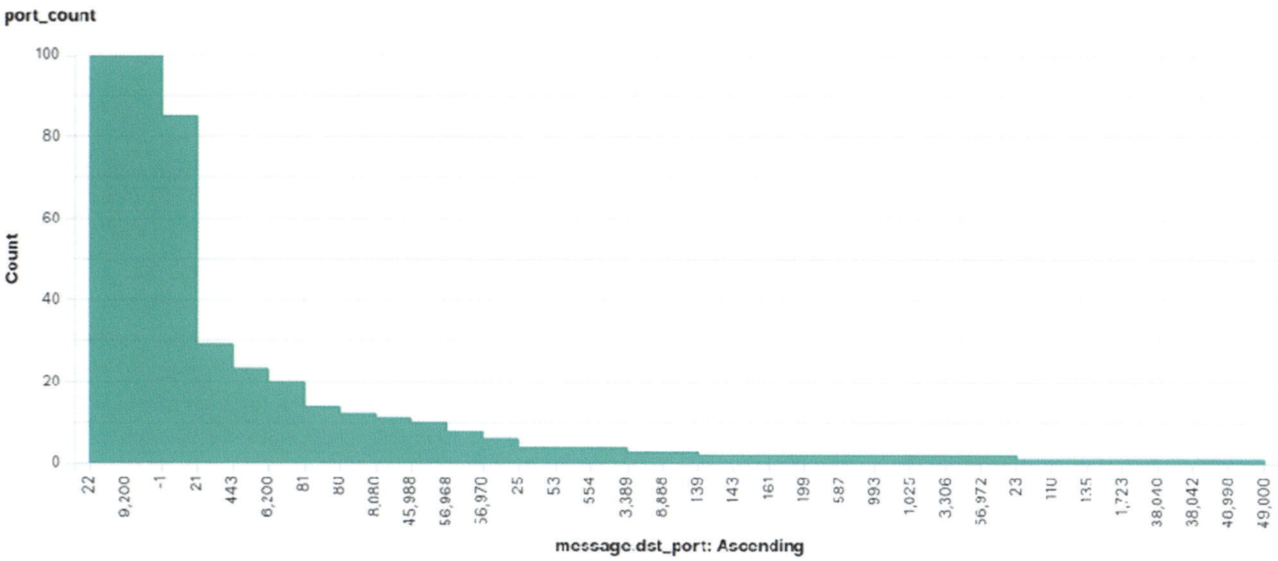

Fig. 6.6 Frequency of ports accessed by attacker

for further analysis to extract information about attackers, types of attacks and the attack pattern etc. The different fields in the log file are as follows:

{"eventid":"cowrie.login.failed",
"username":"honeypots_special",
"timestamp":"2019-05-03T06:29:
36.790718Z", "message": "login attempt
[honeypots_special/hellohoneypots]
failed", "src_ ip" :"172.27.30.60",
"session":"1df889c64361",
"password":"hellohoneypots", "sensor":
"subhasis-BM1AF-BP1AF-BM6AF"}

Cowrie is a medium-interaction honeypot and is useful to trap attackers who use SSH and telnet services.

6.3.3 Clientpot

From earlier studies and literature, we know that attackers try to intrude into client systems by poisoning web servers. We developed a client-side honeypot. It detects attack initiated by malicious or infected web servers on a victim's machine. It is based on a client-server model. The client issues a web URL crawling request to the server and in response server

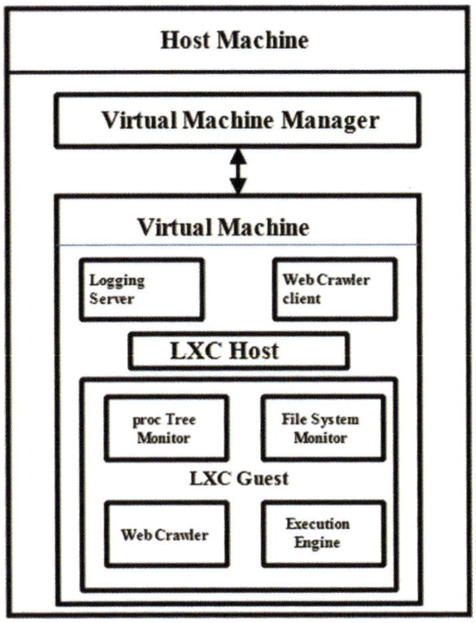

Fig. 6.7 Clientpot framework

```
pid- 980
ppid- 901
url- http://[redacted]
command- lynx http://[redacted]

pid- 982
ppid- 901
url- http://[redacted]
command- lynx http://[redacted]

pid- 986
ppid- 901
url- http://[redacted]
command- wget http://[redacted]
command- pdftk [redacted]
pid 986 KILLED/CRASHED !

pid- 989
ppid- 901
url- http://[redacted]
command- lynx http://[redacted]
```

Fig. 6.8 Log snippet when the malicious web is crawled and malicious pdf is fetched from it

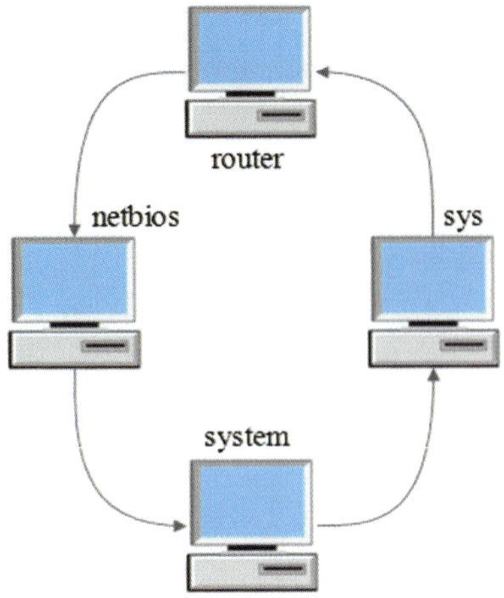

Fig. 6.10 Compute nodes in HoneyFarm and their virtual connectivity

crawls the domain specified by the URL. During the crawling process, if the browser triggers an attack, the system creates a container. The server issues an URL to the container and container processes the downloaded file in a sandbox environment for further analysis. After monitoring the sandbox system, we analyse the logs and decide whether a server is malicious or not. Our model is lightweight in a way that it uses light APIs to mimic Linux-based victim browsers.

Figure 6.7 shows the Clientpot system framework. This system consists of Linux host machine, virtual machine manager (VMM) and Linux Container (LXC) which is free open-source and lightweight software virtualization technology instead of Virtual Machine-based Hardware. VMM manages the virtual machine which is Ubuntu 14.04 LTS. This virtual machine is used to create a sand-boxed environment on the fly. There are two types of Linux containers namely LXC host and LXC guest. Server logging is done at the LXC host, and file and process monitoring are done at the LXC guest.

Fig. 6.9 **a** Logs snippet of Fork Bomb **b** Logs snippet of malicious binary

(a)

```
pid- 845
ppid- 820
url-
command- /bin/sh/fork

pid- 846
ppid- 820
url-
command- /bin/sh/fork

pid- 849
ppid- 845
url-
command- /bin/sh/fork

pid- 851
ppid- 846
url-
command- /bin/sh/fork
```

(b)

```
pid- 843
ppid- 811
url-
command- ~/infected_binary

pid- 846
ppid- 811
url-
command- unzip 23908sadfv98x43560v98.zip
pid 846 KILLED/CRASHED !
```

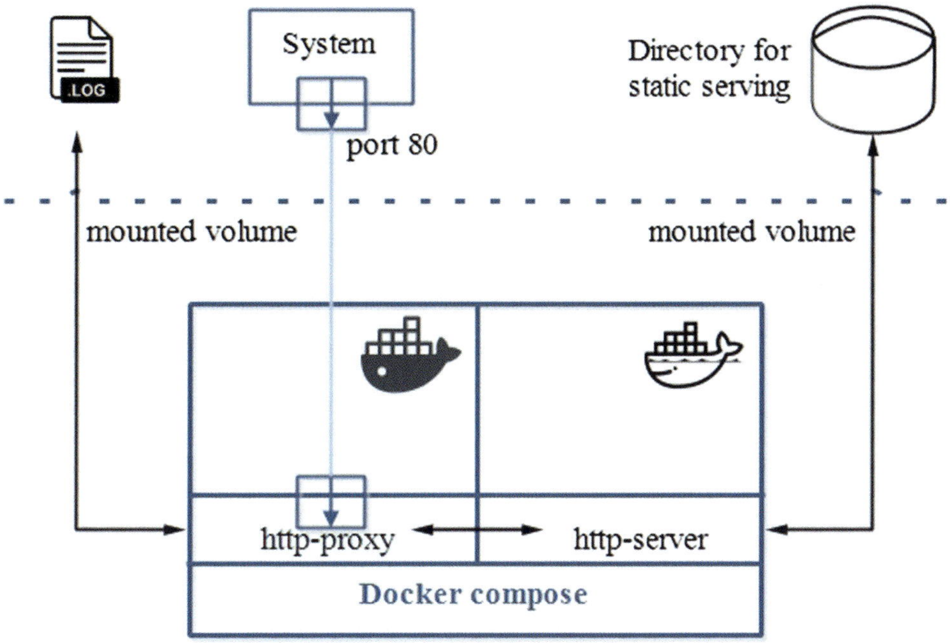

Fig. 6.11 System design of HoneyWEB

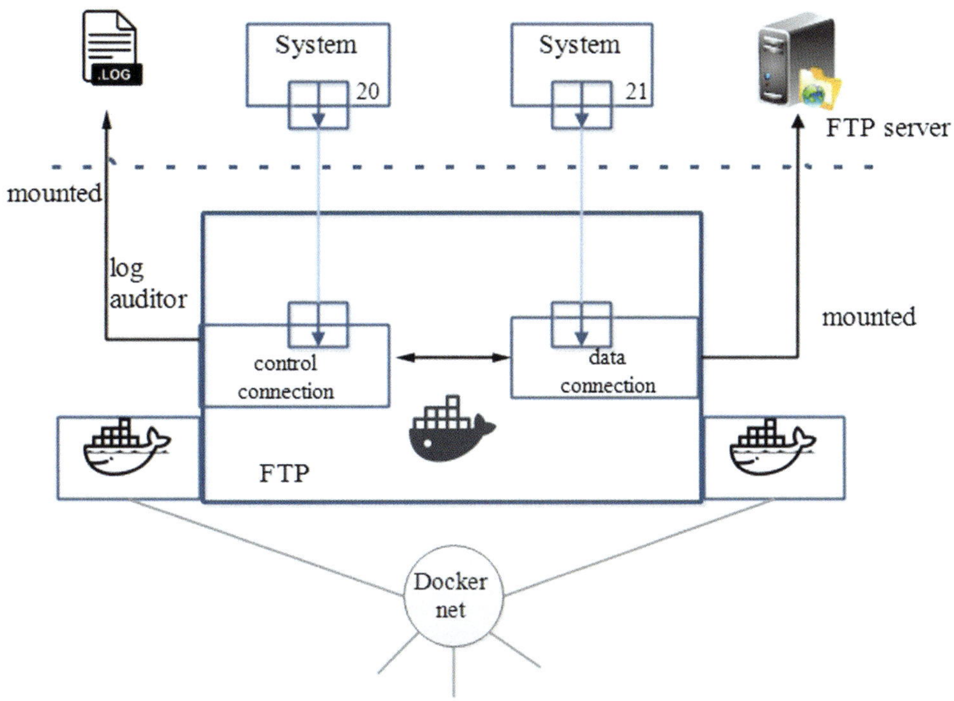

Fig. 6.12 HoneyFTP running in docker network and offering surface for capturing FTP bounce attack

Clientpot provides various services that run on multiple virtual machines. These are `Web Crawler Client`, `Web Crawler Server`, `Execution-Engine`, `File System Monitor`, `Process Tree Monitor`.

Clientpot is tested with several malicious and benign websites. It is able to download suspicious files and analyse them. Figure 6.8 shows the log snippet which is extracted by using a logger service. When URL is crawled, PDF file gets downloaded and crashes the system during execution.

Clientpot model is also used to analyse malware files. So we use clientpot model to analyse two malicious files that are *fork-bombDoS* malware and a *ransomware* and a snippet of these attacks is also presented in Fig. 6.9.

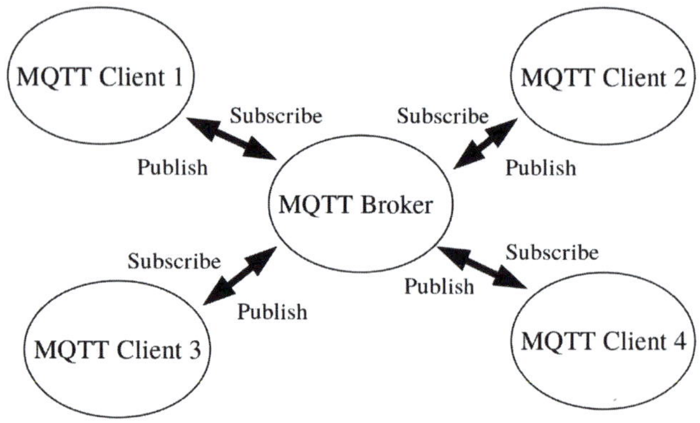

Fig. 6.13 Working of MQTT protocol

6.3.4 HoneyFARM

HoneyFARM is an SSH honeypot which traps the attacker in the network. It is designed in such a way that the attacker spends most of the time in exploring the internal network. When any attacker tries to explore the internal network, HoneyFARM logs the activity and forward the logs to train a Rule-based IDS system. HoneyFARM honeypot is inspired by ssh proxy HonSSH.

HoneyFarm has three major components, namely *open-source SSH proxy, open-source SSH and open-source virtualisation techniques (Linux containers).* Figure 6.10 defines the SSH redirection view of the network from the attacker's viewpoint. The constructed virtual network in the system has four compute nodes. A bridge network connects these nodes. If someone (sys) within the system performs ssh to any other machine across the network, then that person is redirected to a machine named router in the same bridged network. In this way, the attacker is monitored while spending a good amount of time in the virtual network mistaking it as a network of devices.

Fig. 6.14 List of common IPs among honeypots

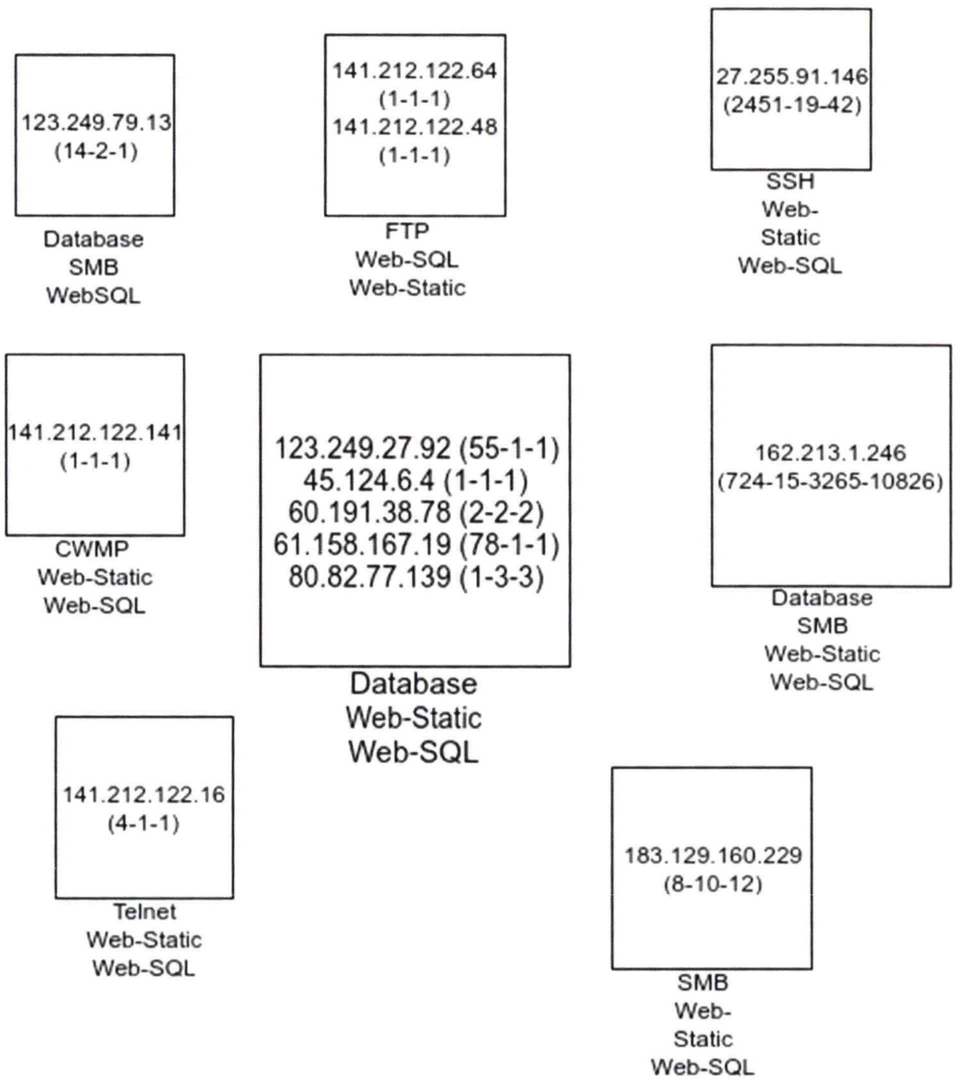

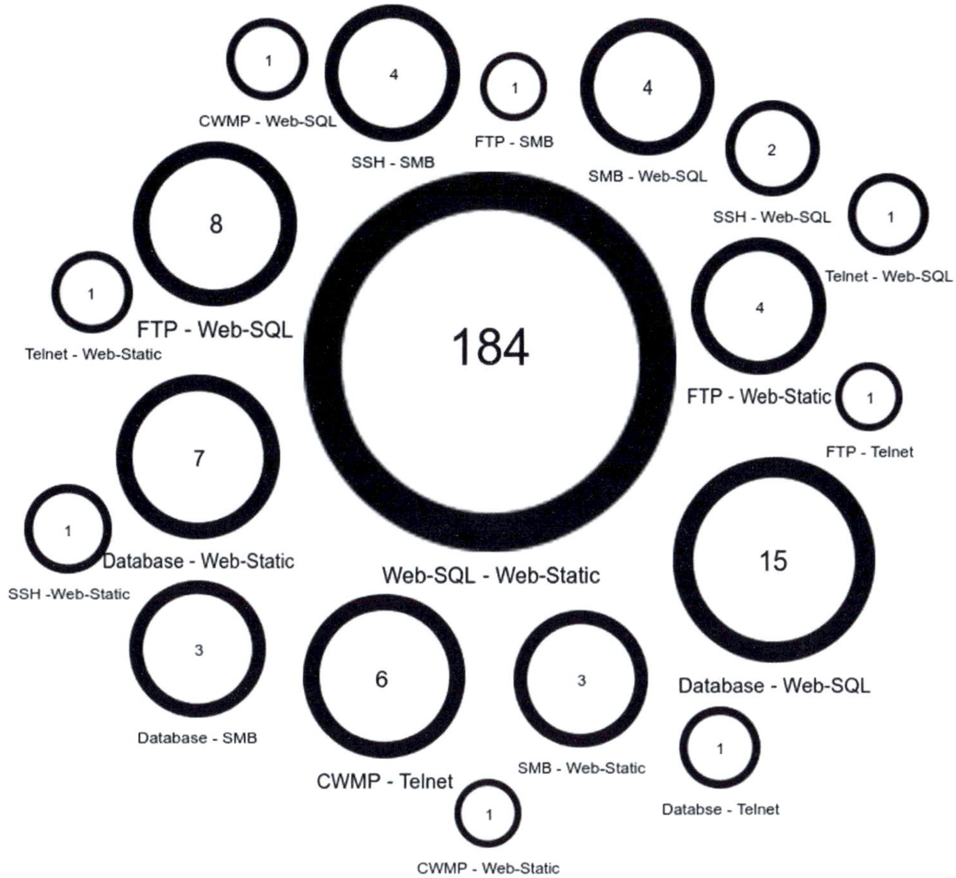

Fig. 6.15 Common IPs among honeypots

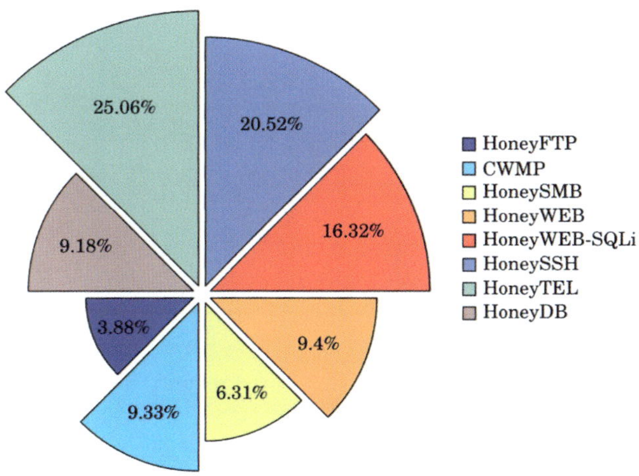

Fig. 6.16 Distribution of observed IPs targeting various honeypots for a duration of 10 days

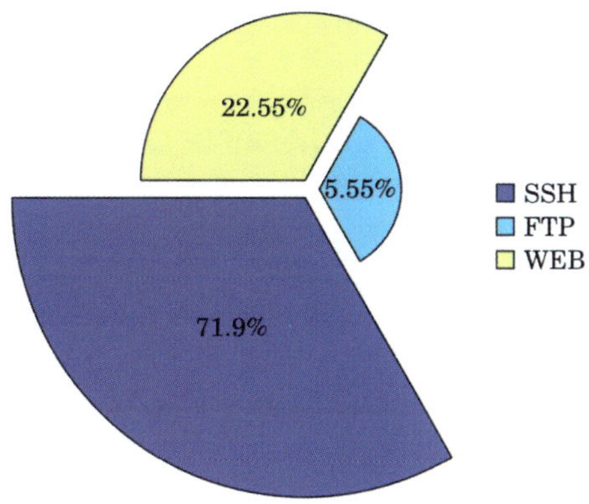

Fig. 6.17 Distribution of observed IPs targeting ssh, web and ftp Honeypots for a duration of 10 days

In this model, there are four virtual systems running in the same physical machine running ssh proxy. When any attack gets executed the following happens:

- An attacker targeting ssh attack on the physical machine is captured by the proxy. This proxy is already configured with a target virtual machine named as system in Fig. 6.10.
- Proxy transfers the request to virtual machine 'system'.
- When the attacker tries to do ssh from 'system' then the attacker is redirected to other replicated devices named 'sys', from 'sys' to router and so on.

Table 6.2 Sample of binary files collected by HoneySSH

Nature of malicious file	No of malicious file
Linux/XorDDos	20
Linux/Bandwidth scripts	13
Linux/Dos	13
Linux/Full Dos Kit	13
Linux/Mirai	3
linux/backdoor	2
Linux/ssh brute-force	2
Windows/Trojan-Ramgo.A	1
Windows/Trojan-Dridex	1
Classified as data files successfully	26

Table 6.3 Number of attempts made by IPs to successfully identify SSH user's password

Number of attempts	IP count
Attempts < 1000	1215
$1000 \leq$ Attempts < 10000	24
Attempts ≥ 10000	7

6.3.5 HoneyWEB

HoneyWEB honeypot is developed to enhance the capabilities offered by the existing web-based honeypot (*glastopf*), which simulates a web server. In HoneyWEB system, the applications are built on the top of a *docker-compose* environment. This creates a network of two different machines connected via *docker-compose* which isolates them from the actual system on which they are running. The proxy opens the WAP port to the system. The interaction of the user takes place via this port and the requests are then logged, manipulated and handed over to another docker which serves the website statically. The logs of the system are mounted to a particular directory in the host of the docker server. Figure 6.11 gives the entire architecture of the model.

In this model, the port 80 of the docker host system is forwarded to the port HTTP exposed port of the HTTP-proxy docker. Then the port 80 of HTTP-server docker is linked via *docker-compose*. When the attacker targets the system for port 80, the following steps happen:

- Attacker sends a malicious link and the requested URL is forwarded to the server and to HTTP-proxy.
- HTTP-proxy logs the request and processes it so that it may not harm the system running in the docker for the HTTP-server.
- HTTP-Server handles the request and sends the response, in return HTTP-proxy checks if the response is `404 Not Found`. If so, it initiates a request for index.html in the HTTP-server.

The HoneyWEB system is built using httrack, docker, docker images, docker-compose, node.js and node.js express. These open-source technologies are integrated in such a way that they offer the entire functionality of the system depicted in Fig. 6.11.

6.3.6 HoneyFTP

HoneyFTP offers a light weight honeypot platform to monitor and analyse various FTP-based attacks and captures malicious executables. Instead of simulating an FTP server like other honeypot solutions do, it creates a docker that actually runs an FTP server. Upon docker startup, the system creates an isolated environment and runs the service that creates two different daemons running on different ports. Figure 6.12 explains the design of HoneyFTP. The service for the FTP server with two different daemons is executed in the docker network within the docker default bridge. The system is made to connect with the two different docker guest machines as well. One of these docker machines allows brute forcing of ssh attacks as well. This invites the attacker to scan the internal network after observing the open ssh port.

For the implementation of the HoneyFTP honeypot, different components are used. They are docker, docker network, *pyftpdlib*—an open-source python library, etc. The docker machine mounts a directory in the host machine, that serves as the base to the FTP directory. In the same machine, two different services are running, one saves the file in the docker host system and other logs the activity of the user on the host.

Table 6.4 Number of attempts made by the IP whose count was ≥10K

IP address	Attempt count
61.177.172.24	12329
61.177.172.17	12680
61.177.172.17	15734
183.214.141.104	27170
183.214.141.104	38128
58.218.211.78	49315
58.218.211.78	71158

Table 6.5 Statistics of HoneySMB

Statistics	Count
No. of unique IP attacked	867
No. of time attacked	1147
No. of malicious files downloaded	21

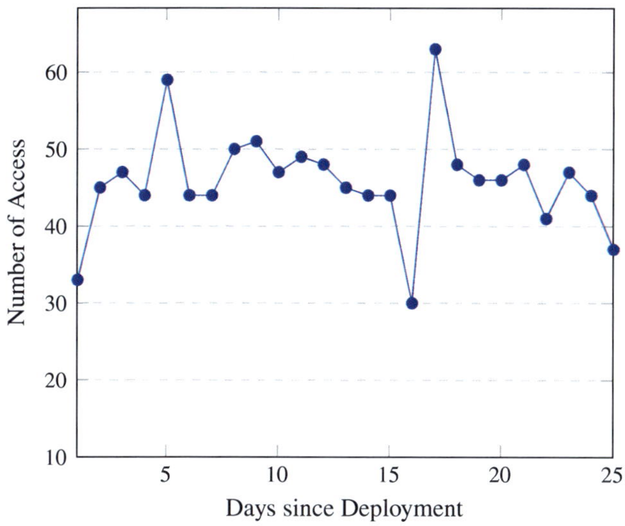

Fig. 6.18 Number of attacks on HoneySMB for different days

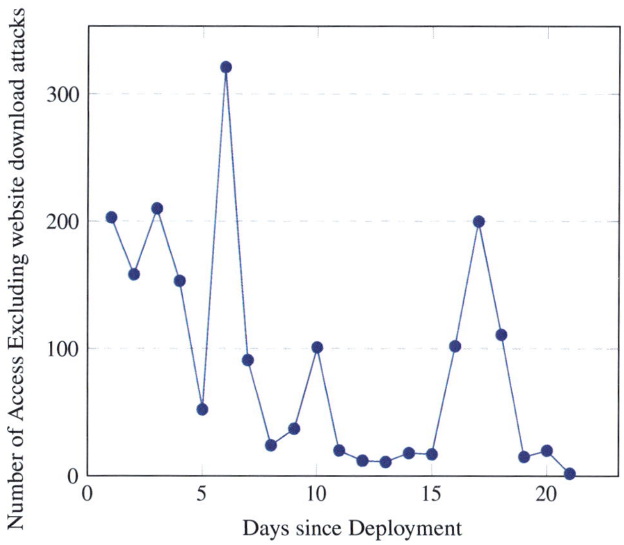

Fig. 6.19 Number of attacks on HoneyWEB for different days

6.3.7 Cloud-Based IoT Honeypot

The design of the cloud-based IoT honeypot is based on ZigBee Honeypot [10] and ThingPot [11]. Message Queuing Telemetry Transport (MQTT) is used as a honeytoken in this honeypot, and MQTT broker is being modified to log all the interaction with the attackers. MQTT is a publish/subscribe messaging standard protocol, which works on top of the TCP/IP protocol to ensure reliable delivery [12] as shown in Fig. 6.13. It helps to minimise network bandwidth and makes it perfect for communication in IoT devices because of it's lightweight feature.

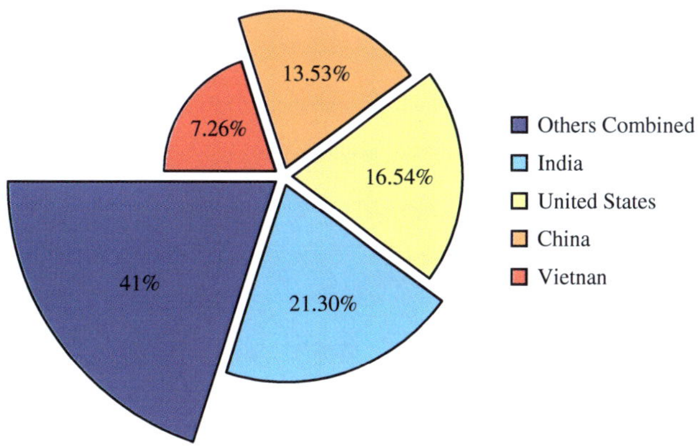

Fig. 6.20 Distribution of attacks from various countries on HoneyWEB

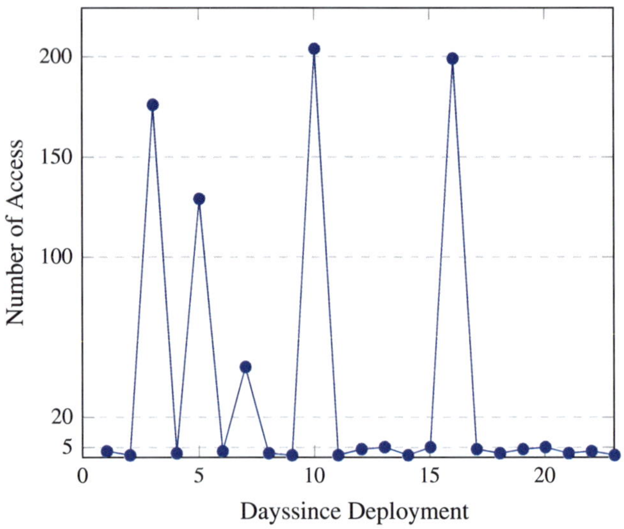

Fig. 6.21 Number of attacks on HoneyFTP for different days

In this Honeypot, we trace the attacks related to the MQTT protocol by putting the MQTT Broker inside the high-interaction SSH Honeypot. Honeytokens, which are IP addresses and port numbers of simulated IoT devices are also placed inside the SSH Honeypot to interact with the MQTT Broker. SSH Honeypot is running in a docker container that is placed on a digital ocean server. The ports 22, 1883 and 8883 of the docker container are exposed to the outside network. The TCP/IP port 1883 is a reserved port and it is used by MQTT, and TCP/IP port 8883 is for using MQTT over SSL. The MQTT Broker processes MQTT packets, so we modify the MQTT Broker to log the information such as IP address, timestamp of the packet, type of the MQTT packet (Connect, subscribe, publish) and content of the packet contained in the MQTT packet. Two log files are placed inside the docker container. One log file is created by the MQTT broker which includes the information about the packets which are exchanged by the client with the MQTT broker. The second log file contains IP and ports on which the simulated IoT devices are running.

In the IoT simulator, we use three IoT devices which act as publishers and five IoT devices that are acting as subscribers. The IoT device names used in the template are similar to those used in daily life such as AC_sensor, CO2_sensor, Chief_monitor, Dean_mobile, Door_sensor, Motion_sensor, Smart_metre, Temp_Sensor, etc. This helps us in meeting the objective of simulating IoT devices which looks legitimate to the attacker.

6.4 Analysis of Attacks on Our Honeypots

This section briefly describes how attackers penetrate the honeysystems comprising of Honeypot, HoneySSH, HoneyWEB and HoneyFTP which are deployed for up to 720 h and what observation we made. Figure 6.14 shows the list of IPs which are common among three or more than three honeypots and Fig. 6.15 indicates the number of attackers' IPs, which are common among various honeypots.

The analysis is done for the last 240 h, which scales down the count of IP addresses. During the last 240 h, there are 13741 unique IP addresses, which are observed and the distribution of attacks across the different honeypots is depicted in Fig. 6.16. Out of these, only 6021 unique IPs are performing attacks on our deployed honeysystems, and distribution is represented in Fig. 6.17.

6.4.1 Analysis of HoneySSH

The total number of malicious binaries captured by HoneySSH is 88. The samples are collected when the honeypot configuration is 'root' as username and password. Table 6.2

Fig. 6.22 Country wise malicious traffic distribution for HoneyFTP

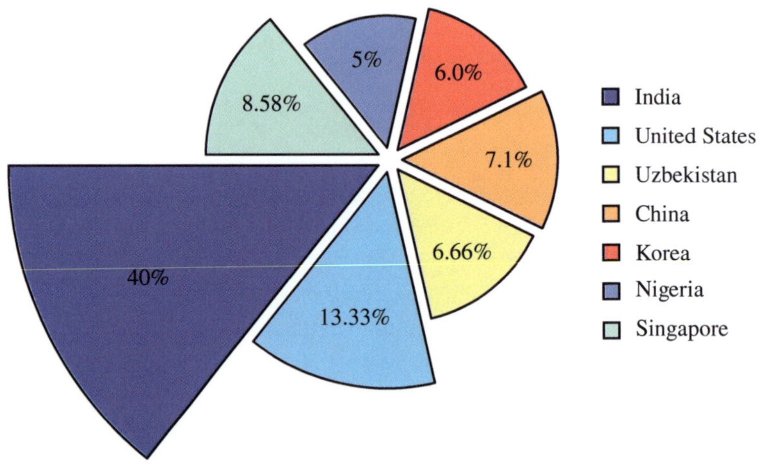

Table 6.6 Top 10 usernames and passwords used during the attacks

Usernames	Number of times used	Passwords	Number of times used
Root	55160	123456	2671
Admin	3573	password	2255
User	1934	123	1769
Test	1901	1234	1283
Ubuntu	1391	12345	1187
Guest	1364	qwerty	1020
Ubnt	1308	p@ssw0rd	1007
Oracle	640	test	894
Postgres	578	123456789	770
Jenkins	392	12345678	751

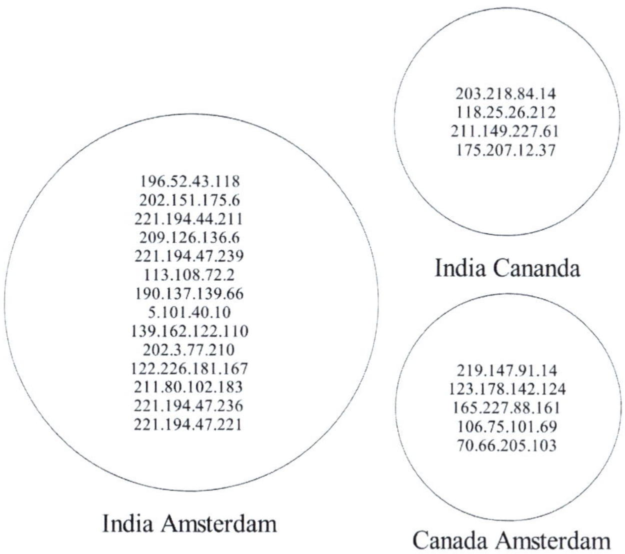

Fig. 6.23 Common IPs between different honeypots deployed over the cloud

shows the number of binaries captured. The total number of the unique IPs which are recorded by HoneySSH system was 1246. The distribution of attempts performed by these 1246 IPs is provided in Table 6.3. Out of 1246 IP addresses, there are 7 IP addresses which made a maximum number of attempts. Table 6.4 shows those 7 IPs and the corresponding attempts.

6.4.2 Analysis of HoneySMB

We deploy HoneySMB for around 25 days across various location in the world. Table 6.5 shows the statistics of HoneySMB. Figure 6.18 shows the number of attacks on HoneySMB per day.

6.4.3 Analysis of HoneyWEB

We collect 568 unique IP addresses, which targeted HoneyWEB. These IP addresses are collected in the first 600 h of uptime. Figure 6.19 depicts the attack trends on different days. We find that HoneyWEB is targeted by 38 different countries, out of which the maximum number of attacks are originated from China-based servers. The distribution of attacks from various countries on HoneyWEB is shown in Fig. 6.20.

6.4.4 Analysis of HoneyFTP

HoneyFTP captures 460 unique IP addresses. The following information is extracted from the attacks performed by these IPs:

1. The attack patterns specified in Fig. 6.21 are observed during the deployment of HoneyFTP. On specific days there is evidence of a sudden increase in attacks in the HoneyFTP.
2. HoneyFTP encountered attacks from 8 geographically distributed locations, and the distribution is shown in Fig. 6.22.

6.4.5 IoT Honeypot Analysis

IoT honeypot is deployed at different locations in India, Amsterdam, and Canada over the cloud for more than a month, and the attacks are received from a total of 2576 unique hosts. Table 6.6 shows top 10 unique usernames used during the attacks.

Figure 6.23 shows the common hosts which target the honeypots among India–Amsterdam, India–Canada and Canada–Amsterdam. There are attacks from IPs—221.194.47.221, 221.194.47.236 and 221.194.47.239, which belongs to the same domain.

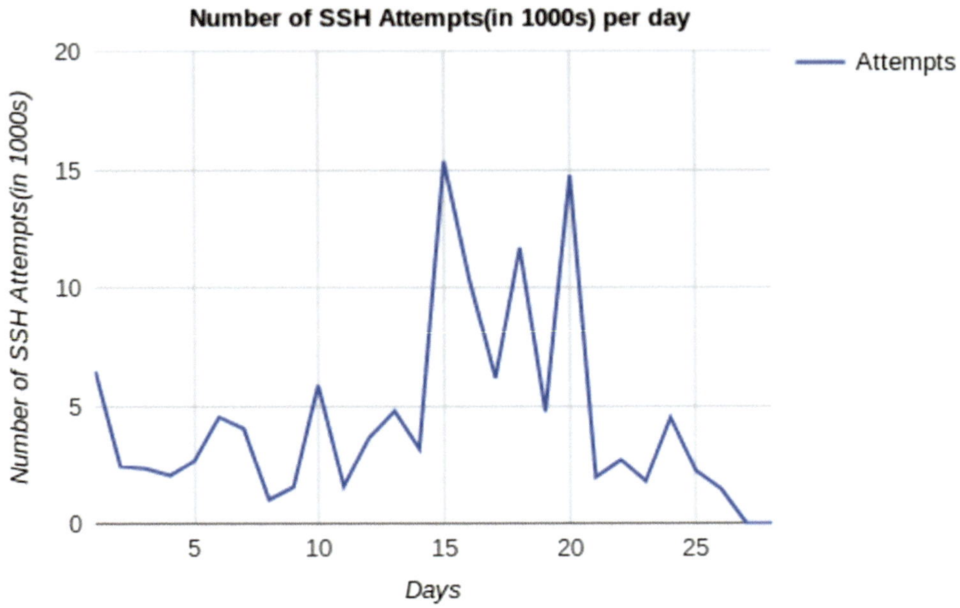

Fig. 6.24 Number of SSH attempts made over a period of 30 days

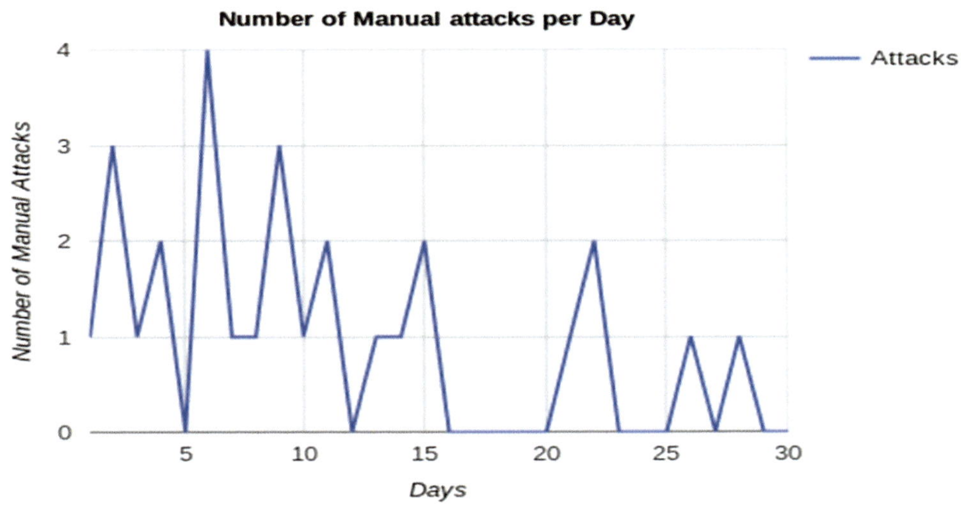

Fig. 6.25 Number of manual attacks per day

Figure 6.24 shows the number of attempts made by the attackers on the deployed honeypot. It shows that initially, the number of attempts increases slowly and then decreases gradually over time. Figure 6.25 indicates the number of manual attacks performed in a time duration of 30 days. Manual attacks are identified by checking the difference between the timestamp of the commands used by the attacker. Usually, automated attacks are made using the scripts in which there is no difference between the timestamp of the commands used by the attacker. Figure 6.26 shows the distribution of IPs which are involved in flooding of connect packets to the broker. Some of the IPs like 116.255.215.93, 117.5.165.188, 103.1.65.85 and 58.87.67.254 are involved in multiple sessions of flooding of connect packets. It is observed that IP 116.255.215.93 is involved in flooding for maximum days. There are more than 1000 connect packets are observed from the IPs as mentioned earlier, which attempts to make Denial of Service attack on the MQTT broker.

6.5 Conclusion

In this chapter, we discuss the implementation and design of various types of honeypots and the difference between honeypot and the traditional techniques like firewall, IDS and IPS. These honeypots use services like SSH, FTP, SMB protocol, SQL-injection, MySQL database, etc. to engage the attacker for a more extended period and generate several logs for further analysis. After the logs analysis, we design the defensive techniques. This chapter also presents the working of the MQTT protocol-based honeypot, which is deployed to generate the attack pattern of IoT devices.

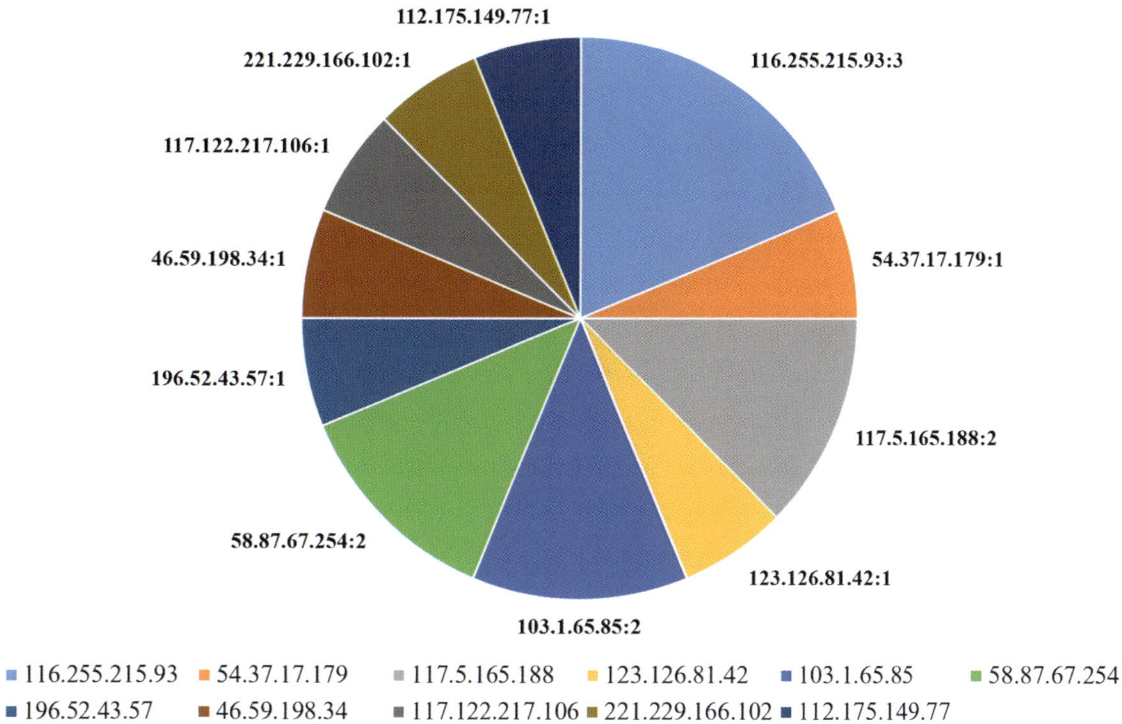

Fig. 6.26 Connect packet flooding IPs

References

1. Team H (2018) Internet stats and facts for 2019. https://hostingfacts.com/internet-facts-stats/. Accessed 30 May 2019
2. Technologies P (2017) Web application attack statistics. https://www.ptsecurity.com/ww-en/analytics/web-application-attack-statistics-q1-2017/. Accessed 29 May 2019
3. Raynal F, Berthier Y, Biondi P, Kaminsky D (2004) Honeypot forensics part 1: analyzing the network. IEEE Secur Priv 2(4):72–78
4. Spitzner L (2003) Honeypots: tracking hackers, vol 1. Addison-Wesley, Reading
5. Yeldi S, Gupta S, Ganacharya T, Doshi S, Bahirat D, Ingle R, Roychowdhary A (2003) Enhancing network intrusion detection system with honeypot. In: TENCON 2003 conference on convergent technologies for Asia-Pacific region, vol 4. IEEE, pp 1521–1526
6. Döring C (2005) Conceptual framework for a honeypot solution. Interactive systems and technologies: the problems of human, p 179
7. OpenCanary honeypot (2017). https://opencanary.readthedocs.io/en/latest/. Accessed Jun 2019
8. Banon S (2010) ELK (Elasticsearch, Logstash, and Kibana). https://www.elastic.co/elk-stack. Accessed Jun 2019
9. Oosterhof M (2015) Cowrie medium interaction honeypot. https://cowrie.readthedocs.io/en/latest/index.html. Accessed May 2019
10. Dowling S, Schukat M, Melvin H (2017) A zigbee honeypot to assess IoT cyberattack behaviour. In: 2017 28th Irish signals and systems conference (ISSC). IEEE, pp 1–6
11. Wang M, Santillan J, Kuipers F (2018) Thingpot: an interactive internet-of-things honeypot. arXiv:1807.04114
12. Hunkeler U, Truong HL, Stanford-Clark, A (2008) MQTT-S-a publish/subscribe protocol for wireless sensor networks. In: 2008 3rd international conference on communication systems software and middleware and workshops (COMSWARE'08). IEEE, pp 791–798

Cache Based Side-Channel Attacks

7

Biswabandan Panda

Abstract
In the current era of computation, multiple processor cores are deployed in devices that range from smart-phones, laptops, desktops, servers, and cloud-based systems. Though, innovations in the world of hardware and computer architecture have resulted in faster computation in terms of better performance, a lot of sensitive data that is stored and processed by these devices can get leaked through various hardware components, such as branch predictors, caches, Translation lookaside buffers (TLBs), hardware prefetchers, Dynamic Random Access Memory (DRAM) controllers, and the DRAM. Basically, these hardware components become side-channels and leak information through attacks known as side-channel attacks. A recent and famous attack that was of similar taste was the famous Spectre and Meltdown attacks. This chapter discusses some of the famous side-channel attacks on on-chip SRAM cache memories and off-chip DRAM memory.

Keywords
Side-channel attacks • Timing channels • Caches • DRAM • Processor

7.1 Introduction to Memory Systems and Side-Channel Attacks

Modern multicore systems have multiple levels of memory to solve the decades-old "memory-wall" problem so that the processor can execute more instructions in fewer cycles, improving the instruction-level parallelism (ILP). Typically, smaller and faster caches are employed closer to the processor, followed by bigger and slower caches and then the off-chip DRAM. Modern multicore systems have per core private L1s and L2s, followed by a shared last-level cache (LLC, also known as L3) that is shared by all the cores. The LLC is usually sliced (banked) based on the number of cores on a chip. To hide/tolerate the off-chip memory latency, caches also hardware prefetchers that prefetch data into the cache before the processor demands for the same, resulting in better system performance. Prefetching is a speculation technique that predicts the future addresses that might be accessed by the processor by looking into the past accesses.

Figure 7.1 shows a logical view of a typical cache hierarchy. Note that there is an on-chip DRAM controller that communicates with the LLC (L3) and communicates the LLC misses to the off-chip DRAM and responds back with the data to the LLC. Recently, most of the multicore systems are SMT/HT based, meaning two hardware threads share the processor core pipeline to execute the instructions. The hardware threads that belong to the same core also share the L1 cache, L2 cache, and other hardware structures such as translation lookaside buffers (TLBs), and hardware prefetchers.

Side-channel Attacks: Side-channel attacks are attacks that lead to information leakage from an application, not because the application is nonsecure, rather because of a channel that leaks information, when the application is executed on a system. These channels exploit attributes like latency, power, and heat. In this chapter, we concentrate on cache-based side-channel attacks that exploit the difference in access latency between a cache hit and a cache miss. At this point, we can make a loose statement that a shared resource that is shared by multiple cores (threads) provides an opportunity for side-channel attacks. In this chapter, we mainly focus on the cache attacks at the different levels of cache, and their mitigation techniques. Primarily, in cache-based attacks [1–12], an attacker accesses its own memory space and times it to infer about the victim's access to the shared resources (shared LLC in case of multicore systems, shared L1/L2s between threads of a hyper-threading core). All the practical attacks can be categorized into hit type/miss type attacks and access-based/timing-based attacks. Table 7.1 shows the same.

B. Panda (✉)
Indian Institute of Technology Kanpur, Kanpur, India
e-mail: biswap@cse.iitk.ac.in

S. K. Shukla and M. Agrawal (eds.), *Cyber Security in India*, IITK Directions 4,
https://doi.org/10.1007/978-981-15-1675-7_7

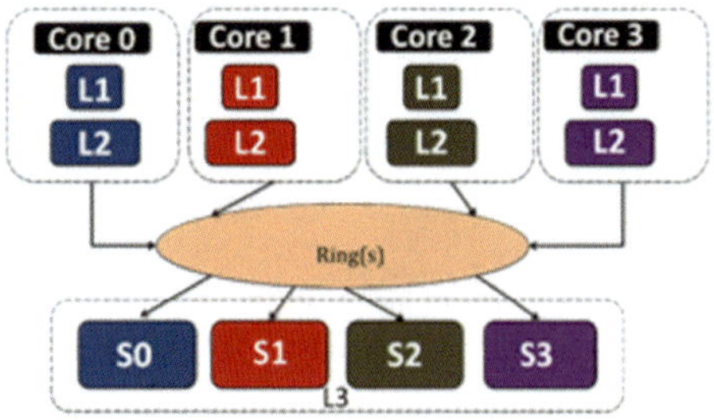

Fig. 7.1 A typical cache hierarchy with private L1s, L2s, and a shared L3 shared by 4 processor cores through a ring interconnect

In miss type attacks, the attacker is interested in observing longer cache access time, because of cache misses (miss access can be either from the victim or the attacker). In contrast, in hit-based attacks, the attacker is interested in shorter access time (hits). All the attacks measure LLC access time. However, some attacks do it precisely per memory access (access-based attacks), and some accumulate the timing information for the entire security-critical accesses (timing-based attacks). Also, all these attacks are possible at different levels of caches: the private L1 and L2, and the shared LLC. Through this chapter, we explain different attack strategies with examples.

7.2 Side-Channel Attacks and Information Leakage

How does side-channel cache attacks leak information? Algorithm 1 shows a pseudocode of an algorithm called square–multiply exponentiation, which is used in many ciphers, where the exponent bits are the secret key that the attacker tries to leak through cache attacks. Note that the key bits can be "0" or "1". If an exponent bit is "1" then there is an additional multiply operation otherwise it is just a square operation.

```
x ← 1;
for i ← |e| − 1 downto 0 do
    x ← x^2 mod n;
    if (e_i = 1);
    then
        x = xb mod n;
    end
    endif;
end
done;
return x;
```

Algorithm 1: Pseudo code of Square-and-multiply exponentiation algorithm.

7.2.1 Attacks of Interest

Evict+Reload: In Evict+Reload attack, the spy core evicts a cache block from the LLC that results in a back-invalidation and invalidates the corresponding cache blocks in private L2 of the victim. After an interval (predetermined fixed value), the spy reloads the same address and if it gets a shorter access time (an LLC hit), then it concludes that the victim has accessed the same cache block. A variation of evict+reload attack is flush+reload [16, 17] attack where the attacker uses clflush instruction to flush the cache line from the entire cache hierarchy and then reloads the same block.

Prime+Probe: In this attack, the attacker loads its cache blocks by evicting the blocks of the victim (the prime part). Then the victim executes its secure operation and in the process, gets LLC misses, evicts the blocks brought by the attacker. Next, the attacker probes its execution time by reloading its blocks, to see whether it gets longer access time because the victim has evicted the block (an LLC miss). Out of all these attacks, Evict+Reload attack demands the notion of sharing of OS pages between the victim and the spy. The attack is more precise (operates at specific block addresses) compared to Evict+Time and Prime+Probe, that operate at the cache set level. Figures 7.2 and 7.3 illustrates one hit type and one miss type attack.

7.2.2 Side-Channel Attacks on Real-World Applications

Evict+Time on AES: In this attack, the attacker attacks a cipher named AES where it evicts an AES cache block containing table entries and then calls a routine to encrypt with random plaintext and measures the encryption time. The attacker will be successful if it gets L1 misses for specific plaintext values, increasing the average encryption time (Fig. 7.4a). In an ideal case, there should not be any difference in execution time as shown in Fig. 7.4b.

Evict+Reload attack on GnuPG and Poppler at the LLC: GnuPG is an open-source implementation of OpenPGP standard. We show the attack on GnuPG 1.4.13 version that uses modular exponentiation in the form of square-and-multiply Algorithm (refer Algorithm 1), where each occurrence of square–modulo–multiply–modulo corresponds to the exponent bit one whereas each occurrence of square–modulo not followed by a multiply operation corresponds to a zero. A complete trace of the square and multiply can help in recovering the exponent. For the sake of simplicity, let us assume an attacker is running on core-0 and GnuPG on core-1. The attacker creates slots where each slot is of thousands of processor cycles in which the attacker evicts the cache blocks of interest and pause and then reload the same cache block at the end of the slot (this pause is required for a successful

Table 7.1 Classification of cache based attacks

	Timing based	Access based
Miss type	Evict+Time [13]	Prime+Probe [14]
Hit type	Collision [15]	Evict+Reload [11]

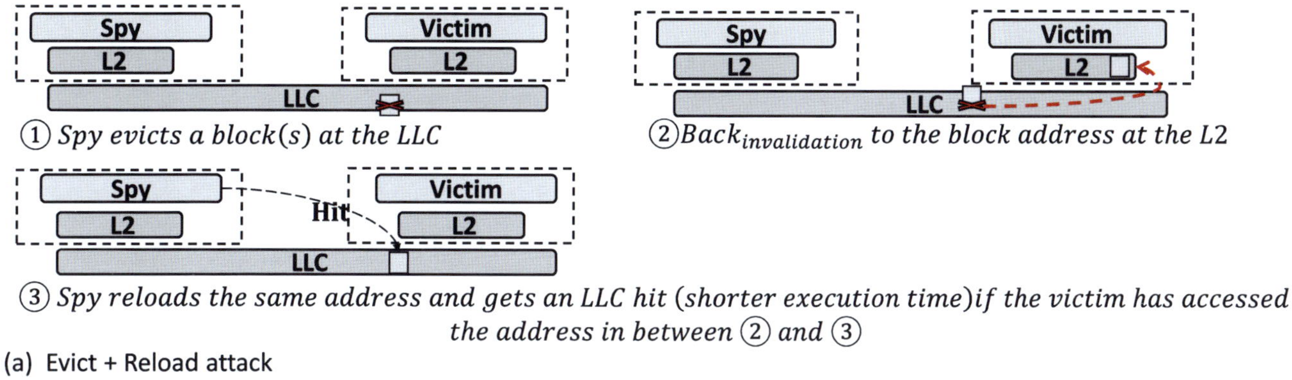

Fig. 7.2 Evict+Reload attack

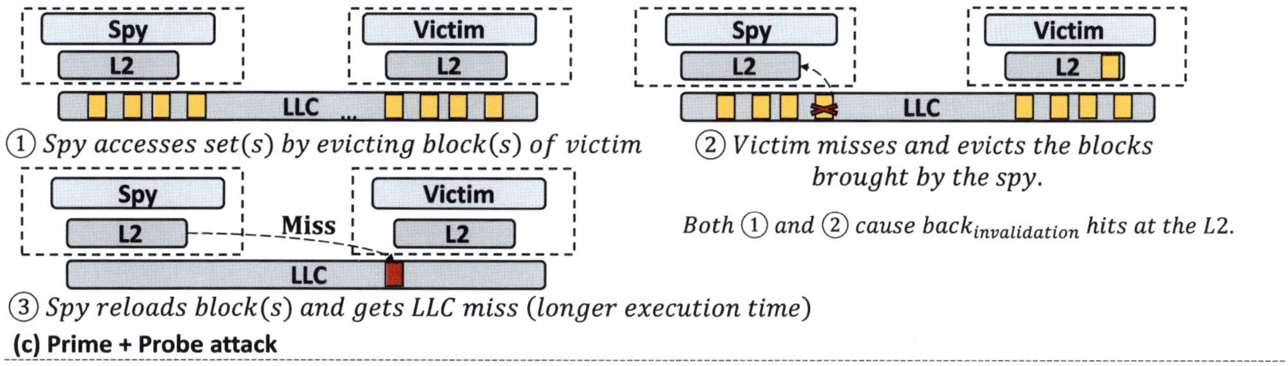

Fig. 7.3 Prime+Probe attack

attack and length of an interval should, therefore, be carefully chosen).

Figure 7.5 shows LLC hits while attacking GnuPG at the LLC for 100 different slots. The LLC hits on square-and-multiply functions correlate to the "0s" and "1s" of the exponent (e) of the Algorithm 1.

7.2.3 What Is Needed for a Successful Attack?

Shared resources play an important role in terms of resource utilization and performance improvement. For a successful cross-core side-channel attack at the LLC, the LLC should be an inclusive one so that eviction of a cache block from the LLC will lead to invalidations at the private L1s and L2s of all the cores. Similarly, cross-thread side-channel attacks are successful because of both the threads of a hyper-threading core share L1 and L2 caches.

For attacks such as flush+reload and evict+reload, there is a demand for sharing of memory space between the attacker and the victim. The sharing can be a simple crypto library that is accessed by both the attacker and the victim. The sharing requirement makes the job of an attacker easy as the attacker knows which particular block addresses to clflush, and which particular cache set to target to evict the blocks of interest, provided the cache indexing bits are reverse-engineered.

Compared to flush+reload and evict+reload, prime+probe attack does not demand any notion of sharing between the attacker and the victim. This makes a strong case for prime+probe attack because even if the victim does not share anything with the victim, the attacker can still mount a successful attack. Also, with the emergence of cloud infrastructure, huge OS pages (sizes more than 4 KB) are becoming common that can help an attacker because the number of bits that get unchanged during the virtual to physical address translation incudes the cache set index, which provides an opportunity for an attacker to know the cache set number.

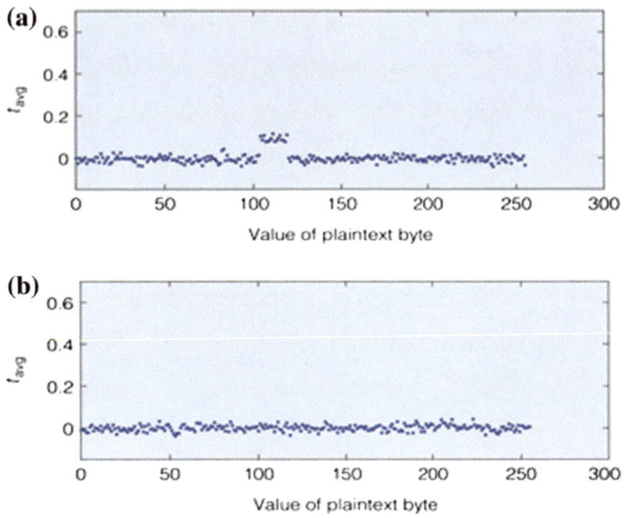

Fig. 7.4 Evict+Time attack on AES. As it is a miss type attack, the attacker will succeed if it gets a relatively higher cache access latencies (execution time) [18]

However, prime+probe attacks take more time as it has to find a set of blocks of interest from a large number of cache sets in the form of an eviction set. Once the eviction set is created, both prime and probe operations are carried out on that set only.

7.2.4 A Case of Spectre and Meltdown Attacks

Spectre and Meltdown [20] are two recently mounted attacks that exploit the speculative and out-of-order nature of the processor. Because of speculative execution, the core brings the data into the cache and once the cache has the data, the existing eviction- and flush-based attacks can be mounted to extract the data. Spectre is more powerful compared to the existing cache attacks as it can leak the actual data as compared to simple eviction-based attacks that can detect the access patterns.

7.3 Countermeasures

Cache partitioning techniques: Cache partitioning techniques like NOMO [21] and DAWG [22] partition the cache in such a way so that the attacker cannot evict the cache blocks of a victim, mitigating the attacks like evict+time, evict+reload, and prime+probe. DAWG partitions the ways of a last-level cache in such a way that the cache replacement and cache hits are restricted to one security domain. Noisy timers: Timewarp [23] is one of the run-time system techniques that try to fudge the timing information (mostly rdtsc instruction) by adding noise. These techniques demand changes at the ISA level, applications level, and for some changes, it needs virtualization support, which is a substantial modification for an architect. These fuzzing techniques do not mitigate attacks that use new techniques (e.g., performance counters) to keep track of micro-architecture events. For applications that need to use rdtsc, these techniques demand changes at the system administrator level. Note that flush-based attacks can be mitigated by preventing clflush instruction in user mode for read-only or executable OS pages (such as shared library code). CEASER [24]: A recent work that mitigates the eviction-based cache attack by using a different cache mapping that makes sure multiple accesses to the same cache block address get mapped different cache block addresses and different cache sets.

7.4 Conclusion

Through this chapter, we discussed the basics of cache-based side-channel attacks, their different forms, their impact, and what has been done so far to mitigate these attacks. Going forward, these attacks, may come in different avatars at other hardware structures like TLBs, prefetchers, and other on-chip shared resources. In the coming future, the systems security community will look for hardening the shared resources and it may come with a performance cost.

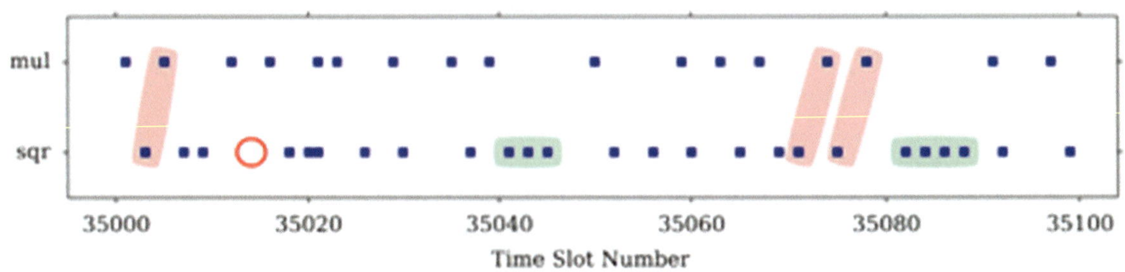

Fig. 7.5 Evict+Reload attack on GnuPG. As it is a hit type attack, the attacker will succeed if it gets an LLC hit (lower cache access latency) [19]

References

1. Neve M, Seifert JP (2006) Advances on access-driven cache attacks on AES. In: International workshop on selected areas in cryptography. Springer, pp 147–162
2. Aciiçmez O (2007) Yet another microarchitectural attack: exploiting i-cache. In: Proceedings of the 2007 ACM workshop on computer security architecture. ACM, pp 11–18
3. Acıiçmez O, Brumley BB, Grabher P (2010) New results on instruction cache attacks. In: International workshop on cryptographic hardware and embedded systems. Springer, pp 110–124
4. Tromer E, Osvik DA, Shamir A (2010) Efficient cache attacks on AES, and countermeasures. J Cryptol 23(1):37–71
5. Gullasch D, Bangerter E, Krenn S (2011) Cache games–bringing access-based cache attacks on AES to practice. In: 2011 IEEE symposium on security and privacy (SP). IEEE, pp 490–505
6. Zhang Y, Juels A, Reiter MK, Ristenpart T (2012) Cross-VM side channels and their use to extract private keys. In: Proceedings of the 2012 ACM conference on computer and communications security. ACM, pp 305–316
7. Benger N, Van de Pol J, Smart NP, Yarom Y (2014) "Ooh aah... just a little bit": a small amount of side channel can go a long way. In: International workshop on cryptographic hardware and embedded systems. Springer, pp 75–92
8. Liu F, Yarom Y, Ge Q, Heiser G, Lee RB (2015) Last-level cache side-channel attacks are practical. In: 2015 IEEE symposium on security and privacy (SP). IEEE, pp 605–622
9. Irazoqui G, Eisenbarth T, Sunar B (2015) S $ A: a shared cache attack that works across cores and defies VM sandboxing–and its application to AES. In: 2015 IEEE symposium on security and privacy (SP). IEEE, pp 591–604
10. Irazoqui G, Inci MS, Eisenbarth T, Sunar B (2014) Wait a minute! a fast, cross-VM attack on AES. In: International workshop on recent advances in intrusion detection. Springer, pp 299–319
11. Apecechea GI, Inci MS, Eisenbarth T, Sunar B (2014) Fine grain cross-VM attacks on xen and vmware are possible! IACR cryptology ePrint archive 2014:248
12. Wang Z, Lee RB (2006) Covert and side channels due to processor architecture. In: 2006 22nd annual computer security applications conference (ACSAC'06). IEEE, pp 473–482
13. Osvik DA, Shamir A, Tromer E (2006) Cache attacks and countermeasures: the case of AES. In: Cryptographers' track at the RSA conference. Springer, pp 1–20
14. Percival C (2005) Cache missing for fun and profit
15. Bonneau J, Mironov I (2006) Cache-collision timing attacks against AES. In: International workshop on cryptographic hardware and embedded systems. Springer, pp 201–215
16. Yarom Y, Falkner K (2014) Flush+reload: a high resolution, low noise, L3 cache side-channel attack. In: USENIX security symposium, vol 1, pp 22–25
17. Yarom Y, Benger N (2014) Recovering openssl ecdsa nonces using the flush+reload cache side-channel attack. IACR cryptology ePrint archive, vol 2014, pp 140
18. Wang Z, Lee RB (2007) New cache designs for thwarting software cache-based side channel attacks. In: ACM SIGARCH computer architecture news, vol 35. ACM, pp 494–505
19. Yan M, Gopireddy B, Shull T, Torrellas J (2017) Secure hierarchy-aware cache replacement policy (sharp): defending against cache-based side channel attacks. In: 2017 ACM/IEEE 44th annual international symposium on computer architecture (ISCA). IEEE, pp 347–360
20. Lipp M, Schwarz M, Gruss D, Prescher T, Haas W, Mangard S, Kocher P, Genkin D, Yarom Y, Hamburg M (2018) Meltdown. arXiv:1801.01207
21. Domnitser L, Jaleel A, Loew J, Abu-Ghazaleh N, Ponomarev D (2012) Non-monopolizable caches: low-complexity mitigation of cache side channel attacks. ACM Trans Archit Code Optim (TACO) 8(4):35
22. Kiriansky V, Lebedev I, Amarasinghe S, Devadas S, Emer J (2018) DAWG: A defense against cache timing attacks in speculative execution processors. In: 2018 51st annual IEEE/ACM international symposium on microarchitecture (MICRO). IEEE, pp 974–987
23. Martin R, Demme J, Sethumadhavan S (2012) Timewarp: rethinking timekeeping and performance monitoring mechanisms to mitigate side-channel attacks. ACM SIGARCH Computer Architecture News 40(3):118–129
24. Qureshi MK (2018) Ceaser: Mitigating conflict-based cache attacks via encrypted-address and remapping. In: 2018 51st annual IEEE/ACM international symposium on microarchitecture (MICRO). IEEE, pp 775–787

Hardware Security in India: The Journey so Far

8

Debdeep Mukhopadhyay

Abstract

Hardware Security is relatively a young discourse when compared to its more classical counterparts, like cryptography or network security. Yet, the growth of this subject in spite of his short history is phenomenal and the impetus of it in the modern-day world is striking. Like other parts of the world, India also joined this important area of research from the last decade to cover important areas in this discourse. This article is a description of some of the core subareas from our country, which includes cryptographic hardware design, side-channel analysis, fault analysis, micro-architectural attacks, Trojan detections, physically unclonable functions, and applications in IoT security. The chapter concludes with some ongoing efforts in the country to extend the core competence developed to develop secured end–end cyber-physical systems, which is of immense importance not only in our country, but also in the rest of the world.

Keywords

VLSI of cryptographic algorithms • Side-channel analysis • Countermeasures • Fault tolerance • Micro-architectural attacks • PUFs • IoT security

8.1 Introduction

Cryptology is the art of making and analyzing algorithms to provide confidentiality, integrity, and availability of information. This encompasses, design of ciphers, and analyzing them via the process of cryptanalysis; developing classical primitives like hash functions, message authentication codes, signature schemes, and the like. Recent days there has been a humongous advance in the field of cryptology, with advanced primitives like attribute-based encryption, functional encryption, fully homomorphic encryption, and last but not the least, post-quantum cryptosystems are being developed and analyzed. These primitives provide additional capabilities, like performing operations on encrypted databases in the pervasive public cloud, yet addressing privacy and security issues. While cryptology deals with the theoretical constructions, practitioners have shown that mathematically strong algorithms are just the beginning. Cryptographic algorithms, like Advanced Encryption Standard (AES), or RSA, Elliptic Curve Cryptography (ECC) when implemented on software or hardware in a naïve manner, there are potential information leakage sources, through what are called as side channels. In the pioneering works of Paul Kocher around 1995, it was reinstated that side-channel sources like timing or power can be utilized effectively to determine the complete secret key using efficient attack algorithms [1]. These side-channel analyses opened up new research directions in cryptographic engineering, which attempts at developing a theory of the reality, which is otherwise not modeled by classical cryptography. In 1997, with a seminal paper from Dan Boneh [2] and followed with works from Biham–Shamir [3], another class of popular attacks, called fault attacks, came to the surface. These attacks showed that popular cryptosystems like RSA or DES can leak the secret keys when there are accidental or unintentional faults in the circuit that computes them. These works show that proper cryptographic engineering does not end with only performance as a criteria, what we also need is protection against side channels and other implementation-driven attacks. At IIT Kharagpur, in order to explore this important and exciting world of cryptographic engineering, the Secured Embedded Architecture Laboratory (SEAL) was set up in 2008. This lab has made fundamental contributions to the space of research in cryptographic engineering, like developing the most efficient fault attack on the worldwide standard AES (Advanced Encryption Standard) cipher, designing fast and small ECC (Elliptic Curve Cryptosystems [4]) hardware IPs, etc. Furthermore, the laboratory has developed the infrastructure and practice to validate the claims through

D. Mukhopadhyay (✉)
Indian Institute of Technology Kharagpur, Kharagpur, India
e-mail: debdeep@cse.iitkgp.ac.in

S. K. Shukla and M. Agrawal (eds.), *Cyber Security in India*, IITK Directions 4,
https://doi.org/10.1007/978-981-15-1675-7_8

rigorous experiments, done on state-of-the-art platform for side-channel analysis.

Hardware security is a field which originates from cryptographic engineering, and applies it to develop a secured hardware layer in computing systems which can serve as a more dependable root-of-trust (RoT) than their software counterparts. Indeed earlier efforts to design the Trusted Platform Modules (TPM) chips, which offer a very low-cost cryptocoprocessor design, typically includes RSA key-generator, SHA-1 hash functions, and encryption–decryption signature schemes. The signature was used for signing and attesting codes, which are considered legitimate to execute. Computers that incorporate a TPM can create cryptographic keys and encrypt them so that they can only be decrypted by the TPM. This process, often called wrapping or binding a key, can help protect the key from disclosure. Each TPM has a master wrapping key, called the storage root key, which is stored within the TPM itself. The private portion of a storage root key or endorsement key that is created in a TPM is never exposed to any other component, software, process, or user. Cryptosystems that store encryption keys directly in the TPM without blinding could be at particular risk adversaries. In 2010, Christopher Tarnovsky presented an attack against TPMs at Black Hat, where he claimed to be able to extract secrets from a single TPM by inserting a probe and spying on an internal bus for the Infineon SLE 66 CL PC [5]. In 2015, reports on differential power analysis attack against TPMs potentially able to extract secrets, were reported [6]. Thus striving to find a hardware security primitive for usage as an RoT is of utmost importance. One of the accompanying challenges is to ensure that the design should be low cost. One promising technology is what are called as Physically Unclonable Functions (PUFs) [7]. The objective of PUFs is to create security primitives which provide every device with a unique fingerprint. PUFs in the black-box sense, receive inputs, which are called as challenges, to result in output response bits. Though there are different technologies on which PUFs can be designed, the most popular ones are silicon based. When the same PUF circuit is implemented on two different devices, the responses obtained for the same challenge to both of them are statistically independent. Thus PUFs offer novel designs of RoTs where the secret key is not explicitly stored in a memory and hence does not need to travel outside the IC boundary. However, PUFs are threatened by powerful machine learning attacks which try to mathematically model these primitives. One of the primary focus of SEAL, IIT KGP is to realize these primitives on hardware platforms, like FPGAs, ASICs, and perform rigorous test on them: with respect to machine learning attacks, reliability analysis with respect to temperature variations, humidity variations, etc.

PUFs are a promising security primitive to address the security concerns in Internet of Things (IoT). IoT products which are commercially being sold and deployed, like surveillance cameras, home automation systems, often have glaring security weaknesses. One of the goals of SEAL is to perform (in)security analysis of these commercially deployed devices. One of the primary reasons why ad hoc authentication mechanisms are often adopted in these devices are high cost of integrating standard cryptographic solutions. The adopted security solutions thus when adopted in the resource-constrained nodes need to minimize processing power, storage space, network bandwidth, energy consumption, and also perform with minimal user intervention. One of the major research goals that we strive is to integrate PUFs into the nodes. However, this also implies designing a new generation of security protocols which are based on the unclonability of PUFs. One of the major challenges for building such protocols is the fact that as PUFs are not clonable, the Challenge–Response Pair (CRP) which the verifier needs is to be explicitly stored. This makes the solution problematic as to store a large database in a secured fashion is cumbersome. One of the way-outs that we make an effort is to combine effectively PUFs with public-key cryptography to have a new generation of PUF-based authentication protocols which can be integrated with IoT subsystems. We aim to build surveillance cameras, smart meters, and other IoT products which have protections against various forms of adversaries like man-in-the-middle attacks, replay attacks due to the usage of PUFs.

IoTs or in general cyber-physical systems are heterogeneous. So, while on one end we have resource-constrained nodes and on other ends there are more powerful machines, servers, and finally the pervasive cloud. In these computing platforms, while side-channel analysis like power analysis, EM analysis may not be directly applicable, there is a very important variant of these attacks, which are called as micro-architectural attacks. These attacks target the underlying computer architecture, which has been fundamentally developed with performance in mind, and show that there are grave security flaws. Timing analysis, vulnerabilities due to the presence of cache memory, branch prediction attacks, rowhammer bugs of DRAM chips are some of the well-known micro-architectural attacks. At SEAL, we have been working on developing these attacks and exploring them on all well-known machines, like from Intel/ARM/AMD and have developed the expertise of not only demonstrating these attacks, but also developed coding practices for being secured against these attacks. These countermeasures are often costly and thus needs to be suitably designed to ensure a minimal footprint on performance. Another related research direction, which we pursue is to develop reactive measures by developing smart monitors in computing systems which can raise early alarms in the system on the presence of side-channel analysis. In this context, we show the use of machine learning based analysis for the detection of various micro-architectural side-channel attacks, like cache misses, branch-misses and even the newer Spectre and Meltdown.

Finally, no discussion on security is complete without bringing in the aspect of cloud. Clouds provide a massively powerful distributed computing paradigm, where service can be rendered and after usage can be relinquished. While on one hand cloud provides many opportunities for high-performance computing and is often conceptualized as the endpoint of the IoT, which can store and process billions of data which is being generated by the sensors, devices in IoT, they also raise grave concerns on security and privacy. Solutions to encrypt them, naturally raise questions on how to operate on encrypted databases. While theoretically feasible solutions like Fully Homomorphic Encryption (FHE) exist, which can serve as the holy grail in giving capabilities of performing arbitrary computations on encrypted data, but they are not feasible. As of now, they do not scale well and make operations like search extremely slow, virtually to the point of being unusable. At SEAL, we focus to develop on one hand new solutions for restricted but useful database computations, like search, using lightweight primitives, like pseudorandom permutations; on the other hand, we strive to leverage the parallelism offered by hardware designs to develop fast accelerators for such search algorithms.

In this article, we give an overview of the above topics, which are some of the highlights of contribution of hardware security research from our country. In particular, we provide the following descriptions which are some of the major hallmarks in hardware security research in India:

- Differential Fault Analysis (DFA)
- Hardware Design of Public-Key Cryptosystems
- PUFs: Design and Usage in IoT Security
- Microarchitectural Threats: Attacks, Countermeasures, and Detections
- Hardware Security to Accelerate Cloud Cryptosystems.

Organization: Sect. 8.2 of the chapter presents an overview of our work on differential fault attacks of cryptosystems, along with research in automating fault analysis. Chapter 3 discusses on VLSI design of public-key algorithms, with ECC in perspective. The chapter gives design approaches developed to cater to both design alternatives, viz. high-performance and lightweight applications. Subsequently, Chap. 4 focuses on PUF designs, briefing on works done on a promising PUF topology which combines machine learning robustness with reliability. It also presents a protocol suited for IoT applications using PUF-based primitives. Chapter 5 presents research efforts in the exciting confluence of computer architecture and security, annotating our findings in the topic of cache attacks, branch prediction attacks, and row-hammer attacks. We also discuss machine learning based detection methodologies for such side-channel attacks as an alternative for costly countermeasures. Finally, Chap. 6 presents a new line of research initiated in our laboratory to develop hardware-based accelerators for novel search algorithms on encrypted data. The overall direction of our research is summarized in Sect. 8.7.

8.2 Fault Analysis of Cryptosystems

Fault attacks are one of the most potent threats to modern secured systems. Given the fact that precise and targeted faults can be induced in most of the modern computing systems by means of commercially available low-cost equipment, fault attacks become reasonably practical. The SEAL Lab at IIT Kharagpur possesses several setups for targeted fault injection using intentional glitches in the clock network, electromagnetic radiation, and laser rays. The research team has successfully demonstrated injections for software and hardware implementations of crypto-algorithms. Practical realization of remote fault injections using the row-hammer vulnerability has also been demonstrated by the team. Evaluation of cryptographic primitives against fault attacks is an active area of research. The general philosophy for key extraction is analyzing both the correct and the faulty outputs from the target primitive. However, the analyses are mathematically intensive and widely vary among different primitives. This article will focus on fault attacks on block ciphers which are, undoubtedly, the most widely deployed cryptographic primitive so far. More specifically, we shall focus on a specific class of attacks called Differential Fault Analysis (DFA), which is the most widely explored and general class of fault analysis.[1] The fault analyses vary depending on the underlying structures of the block ciphers. However, the minimal requirement for an attack is the availability of at least one faulty ciphertext and its corresponding correct ciphertext.

8.2.1 Attacks and Countermeasures

Most of the early efforts of block cipher fault analysis were targeted toward the AES algorithm, which is the current worldwide standard for symmetric key encryption. AES is a Substitution-Permutation-Network (SPN) structure having 10, 12, or 14 iterative rounds, three different key sizes of 128, 192, and 256 bits, and three different state sizes of same lengths with the key. Each of the rounds consists of four bijective Boolean functions defined over $GF(2^8)$, among which one is nonlinear (`SubBytes`), and rest are linear (`ShiftRows`, `MixColumns`, and `AddRoundKey`). One of the major challenges at the initial phase of fault attack research was to evaluate how easily and practically one can extract the complete secret key of AES. One of the seminal works

[1]Other classes of fault attacks use certain target-specific physical assumptions over the nature of the faults. Their analyses are different but heavily influenced by DFA.

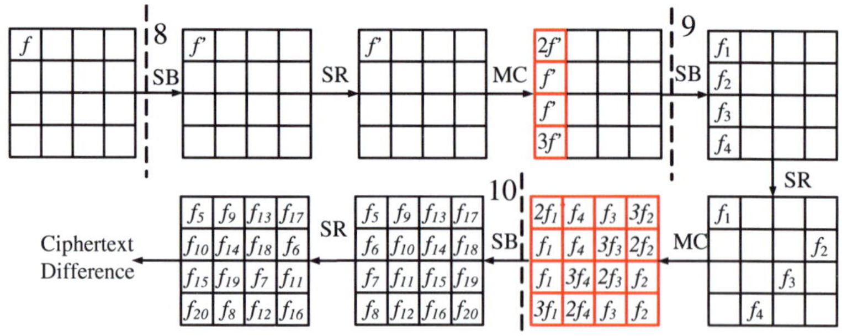

Fig. 8.1 Fault propagation in AES with the fault injected at the beginning of 8th round. The intermediate states which are utilized for constructing equations are marked in red

in this context was due to Tunstall and Mukhopadhyay [8], which showed that one single fault corrupting a byte of the intermediate state of AES at the beginning of 8th round, can reduce the size of the keyspace from 2^{128} to 2^8. As a result, the secret key can be compromised within minutes, even with nominal computational power. Furthermore, this attack has been proven to be optimal from information-theoretical perspectives. Several practical implementations of this specific attack, both on hardware and software AES implementations have been presented in the literature.

To provide a basic understanding of the fault analyses, here we present a brief description of the aforementioned attack. Let us assume a byte fault has been injected at the intermediate state of AES at the beginning of the 8th round. The value of the fault is unknown to the attacker, who can only observe the correct and the faulty ciphertext and guess keys. Each intermediate state, by convention, is represented as a 4×4 matrix of 16 bytes. Without loss of generality, we assume the 0th byte in the matrix to be corrupted. The XOR differential of the correct and faulty states are considered for analysis. As shown in Fig. 8.1, the fault propagates to the ciphertext, eventually corrupting the complete state of the cipher. Here, we have shown the differential propagation of the fault for convenience.

Referring to the differential propagation of the fault in Fig. 8.1, it can be observed that the state at the input of 10th round has certain observable linear patterns in its columns. More specifically, each of the columns is linearly dependent and spanned by a single variable. Exploiting these patterns four independent nonlinear system of equations (over GF(2^8)) can be constructed over associated keys, ciphertext differentials, and fault variables. One of these systems is shown in Eq. (8.1). Here each C_i denotes a byte from the ciphertext, and each k_i denotes the associated key byte. The faults are represented with the variable f_1.

$$
\begin{aligned}
2f_1 &= S^{-1}(C_1 \oplus k_1) \oplus S^{-1}(C_1 \oplus \delta_1 \oplus k_1) \\
f_1 &= S^{-1}(C_{14} \oplus k_{14}) \oplus S^{-1}(C_{14} \oplus \delta_{14} \oplus k_{14}) \\
f_1 &= S^{-1}(C_{11} \oplus k_{11}) \oplus S^{-1}(C_{11} \oplus \delta_{11} \oplus k_{11}) \\
3f_1 &= S^{-1}(C_9 \oplus k_9) \oplus S^{-1}(C_9 \oplus \delta_9 \oplus k_9)
\end{aligned} \quad (8.1)
$$

Solving this system for the 4 associated key bytes is supposed to filter out total 2^8 key choices. Solving all four such systems will reduce the entire keyspace from 2^{128} to 2^{32}. Further, a similar equation system can be constructed for the first column of the 9th round input state differential. This equation system will finally reduce the keyspace to a size of 2^8. The attack will remain exactly the same if any other byte of 8th round input gets corrupted. The attack still works even if the exact location of the corrupted byte at 8th round is unknown to the attacker. The keyspace size, in that case, becomes 2^{12}.

Fault attacks are inevitable for most of the crypto-primitives present today. However, there is a consistent effort for building effective countermeasures against fault attacks. A large class of fault attack countermeasures performs some redundant computation to detect the presence of an injected fault. Redundancy can be realized via multiple encryptions over the same data at the simplest case, or via certain error correcting codes. However, it is not impossible for a clever attacker to bypass such redundant computations either by injecting the same faults in all computation branches or by finding gaps in the error detection capabilities of the codes used. One practical remedy to such advanced attacks is to make the computations randomized by cleverly introducing dummy rounds so that it becomes difficult for the attacker to identify the exact injection location. Moreover, explicit checks between the actual and redundant computations can be avoided. These facts give rise to an entirely different class of countermeasures called infective countermeasures. Infective countermeasures avoid explicit checks and randomize the output upon detection of a fault. To further enhance the randomization it performs dummy round computations between actual and redundant cipher rounds. An efficient infective countermeasure algorithm has been developed by Tupsamudre, Bisht, and Mukhopadhyay in [9].

8.2.2 Automated Detection of Fault Attacks

As already pointed out at the beginning of this section, the attack techniques vary significantly among different ciphers. Even for one block cipher, the entire analysis will vary depending on the location and nature of the fault. Due to such

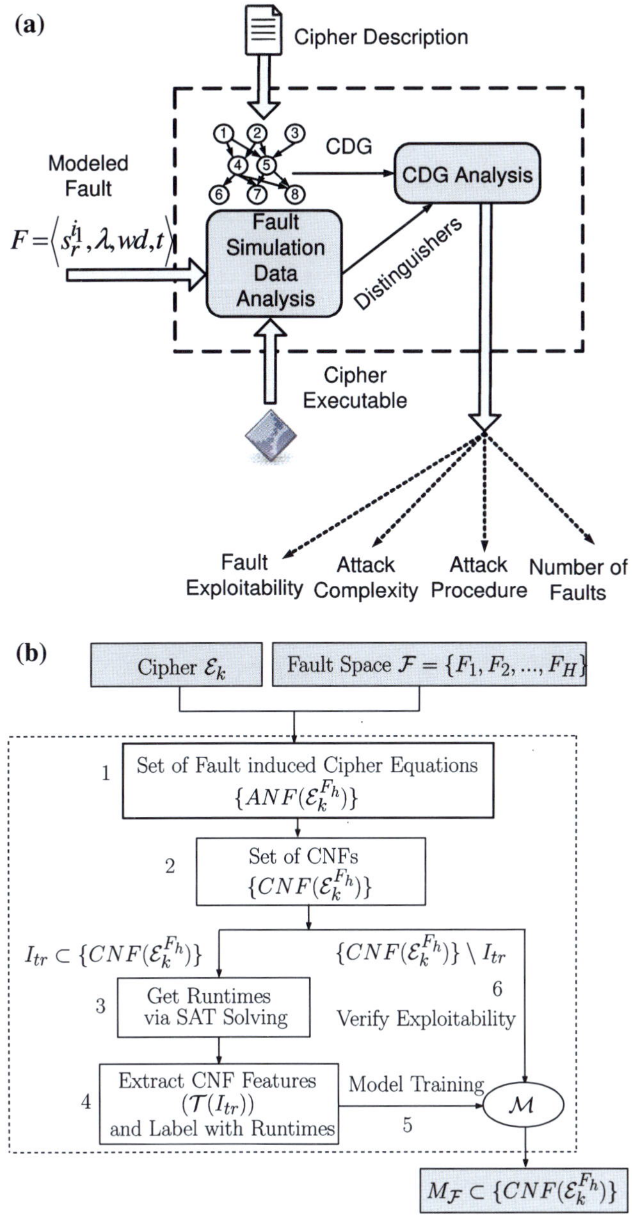

Fig. 8.2 Two automations **a** ExpFault framework [10]; **b** ML-Fault framework [11]

variability, fault attacks are nontrivial to generalize. However, it is not impossible as we shall show in this subsection. The availability of hundreds of block cipher designs makes the generalization and automation of fault attacks essential. Further, the fault spaces for each of these ciphers are prohibitively large in size. Any automation for fault attacks must be fast enough to cover the entire (or at least a significant share) fault space of a cipher.

The three expected properties of a fault attack automation are genericness, speed, and interpretability. An automation has been proposed in [10], which satisfies all these criteria simultaneously. Instead of performing the attacks explicitly, the automation proposed in [10] returns the attack algorithm and its complexity given the specification of a block cipher and a fault model. The necessary criteria behind every fault attack is a wrong key distinguisher, which can filter out a large part of the keyspace thus reducing the complexity of an exhaustive search. As a concrete example of a distinguisher, we refer to the 10th round input differential state in Fig. 8.1. The concept of distinguishers has been formalized in [10] based on Shannon Entropy. Based on the formalization, a fault simulation-based framework has been proposed, which mines out distinguishers from fault simulation data. The next step is to use a graph-based abstraction of the cipher, known as Cipher Dependency Graph (CDG) to figure out the exact attack algorithm (if any) and its complexity. A schematic of this framework, which is called as *ExpFault* is presented in Fig. 8.2a. ExpFault successfully figured out all previously proposed attacks on AES, and PRESENT block ciphers within minutes. Moreover, it figured out optimal attacks on a completely new cipher design called GIFT. Table 8.1 presents a summary of the attacks figured out for the GIFT cipher.

Although ExpFault is able to figure out the fault attacks successfully, it utilizes certain abstractions for the sake of scalability. For example, the internal structures of the S-Boxes are not considered explicitly in ExpFault. Also, it does not explicitly consider the fault values and plaintext values while sketching the attack path. Although these two parameters are not necessary to construct the attack paths in DFA, for certain cipher designs they may influence the attack complexities to some extent. As a result of these approximations, ExpFault returns the best possible attack complexity from the perspective of an attacker. While this information is often sufficient from a cipher evaluator's perspective, it cannot completely justify how an attack may perform on average on a given cipher design.

The only way to handle the above-mentioned issue is to get rid of all abstractions made in ExpFault. It is, however, feasible by means of an algebraic representation of the cipher. Algebraic representation converts each constituent Boolean function of the cipher (and the faults) to a system of nonlinear polynomial equations over $GF(2)$. This system then can be converted to Conjunctive Normal Form (CNF) and solved by Boolean Satisfiability (SAT) solvers. The attack is considered successful if the SAT solver returns the key within a reasonable time. The solver runs forever if the fault is not exploitable. Although SAT solving based approach is fairly generic, it is not fast enough for evaluating each fault. Further, the attacks are not interpretable. To make certain quantitative decisions on the fault space of a cipher, one thus require to make this approach scalable. To improve the scalability, a machine learning (ML) assisted framework is proposed in [11]. The ML engine is able to determine whether a SAT instance is solvable within a reasonable time just from the structure of the CNF representation. As a result, the fault

Table 8.1 Summary of DFA attacks on GIFT. We consider a fault injection attempt a successful attack only if both the evaluation complexity and remaining keyspace size is less than the size of the actual keyspace

Fault width	Round	Attack results				
		Evaluation complexity	$\vert\mathcal{R}\vert$	No. faults per location	Keys extracted	Comments
4	24	—	—	—	—	No attack found
	25, 23	$2^{17.53}$	$2^{7.06}$	1	128	Best attack found
	26, 24	2^{6}	$2^{3.53}$	1	104	Cannot extract full key
	27, 25	2^{6}	$2^{3.53}$	1	72	Cannot extract full key
8	24	—	—	—	—	No attack found
	25, 23	$2^{17.53}$	$2^{7.06}$	1	128	Best attack found
	26, 24	2^{6}	$2^{3.53}$	1	104	Cannot extract full key
	27, 25	2^{6}	$2^{3.53}$	1	72	Cannot extract full key

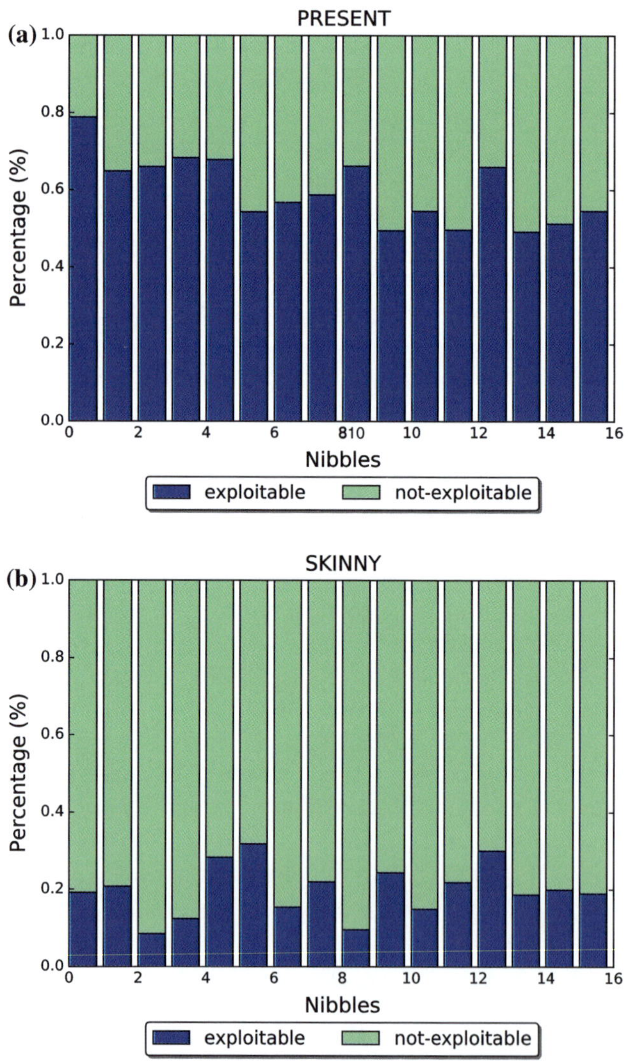

Fig. 8.3 Exploitable fault spaces with **a** PRESENT S-Box; and **b** SKINNY S-Box [11]

characterization becomes extremely fast after a certain number of fault instances are exhaustively characterized with SAT for training. This brings the ability to characterize a large sample of faults within a reasonable time. A schematic of this framework is given in Fig. 8.2b. Several interesting observations can be from these large characterized fault samples.

To illustrate the importance of this fault characterization framework, we present an example adapted from [11]. In this example, the PRESENT block cipher structure is instantiated with two different S-Boxes having similar mathematical properties. Figure 8.3 presents the characterized fault space for these two instantiations corresponding to a specific fault location. It can be observed that the percentage of successful attack instances are significantly different in these two cases, depending on the value of the injected faults. This observation leads to the discovery of certain mathematical properties of the S-Boxes, which were not previously known to have any effect on fault attacks (see [11] for further details). From another perspective, these two graphs represent the average success rate of an attacker for a given fault model, which cannot be trivially obtained from ExpFault at its current stage. However, this ML-based framework cannot provide exact complexity figures and, more importantly, the attack paths. So both the frameworks are somewhat complementary.

8.3 Hardware Design of Public-Key Cryptosystems

Public-key cryptography plays a very important part in a secure communication network. It ensures confidentiality, integrity, and authenticity of the information being exchanged and performs the secure exchange of secret keys for execution of symmetric-key ciphers. Among the available public-key

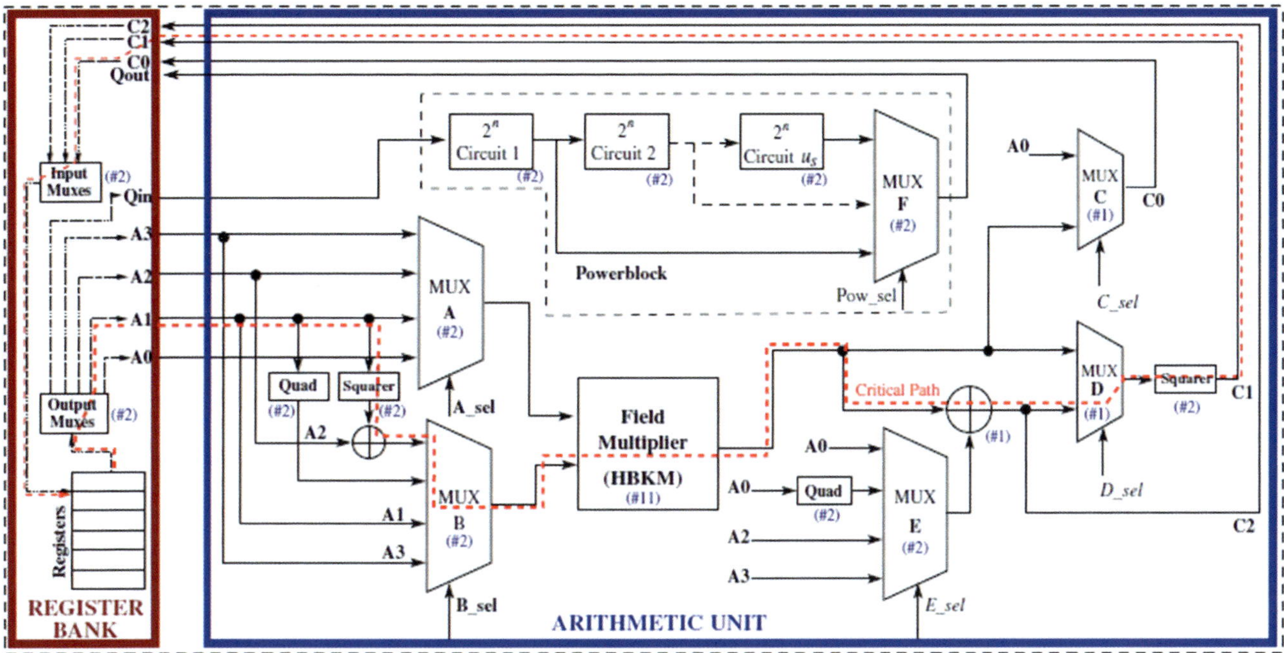

Fig. 8.4 Block diagram of the proposed ECC scalar multiplier architecture in $GF(2^n)$ [13]

cryptosystems, RSA and Elliptic Curve Cryptography (ECC) are the most prominent. The mathematical foundation of ECC is based on the intractability of discrete logarithm problem in elliptic curves. Compared to RSA, ECC provides more security per key bit. However, the computation steps involved in the execution of ECC are mathematically intensive, making its software implementation inefficient and unsuitable for many real-world applications. An alternative approach is to provide dedicated crypto-accelerator on hardware platforms like ASIC (Application-Specific Integrated Circuit) and FPGAs (Field Programmable Gate Arrays) to accelerate ECC scalar multiplication [12]. The execution of ECC is generally carried out in either $GF(2^n)$ or in $GF(p)$. In the SEAL Lab, we have developed efficient implementations for ECC-based cryptography for both $GF(2^n)$ and $GF(p)$ for FPGA platforms which we have summarized in subsequent sections.

8.3.1 Fast and Efficient Implementation of $GF(2^n)$ ECC Scalar Multiplication on FPGA

The most important operation for execution of ECC is elliptic curve scalar multiplication. A typical execution of ECC scalar multiplication takes a point on the curve P and a scalar k as input and produces $[k]P$ as output. In this subsection, we will discuss our proposed high-speed implementation of ECC scalar multiplication in $GF(2^n)$ [13,14]. The architecture of the proposed implementation is shown in Fig. 8.4. The entire architecture for executing ECC scalar multiplication in $GF(2^{163})$ consumes around 148 registers, 10195 lookup

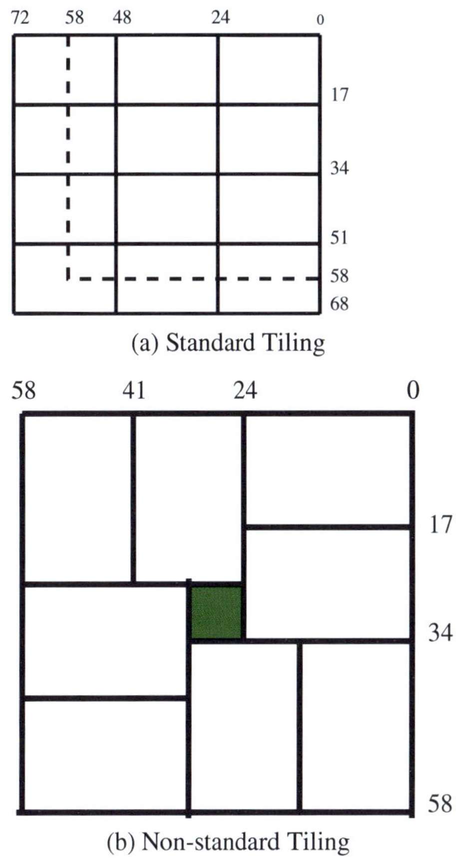

Fig. 8.5 Multiplying operands of width 58 using asymmetric multipliers [15]

tables (LUTs), and 3513 logic slices on a Virtex-5 FPGA. The corresponding latency of the design is only 9.5 μs. This performance escalation is achieved due to multiple architectural optimization techniques that we have incorporated in the design. Few of them are discussed below:

- **Increasing clock frequency with optimal pipelining:** The performance of any hardware implementation is highly dependent on the critical path of the architecture. Pipelining is a common approach which is incorporated in the design to reduce the critical path so that the design can support high frequency of operation. However, pipelining also increases clock cycle requirement of the design. It has been found that blindly applying pipelining often deteriorates the performance of a design as reduction of critical path delay is nullified and usually overshadowed by the increment in the clock cycle requirement. In this work, we have modeled the delay of different mathematical operations like field adders, field multipliers, exponentiation modules. These delay values allow us to partition the critical path in an optimal manner so that the advantage of the pipelining technique is truly assimilated in the architecture.
- **Reducing clock cycle requirement:** Another strategy to improve the performance of the design is to reduce the clock cycle requirement of the architecture. To achieve this, we have first developed efficient and fast implementation of field multiplier and exponentiation module in $GF(2^n)$. The field multiplier architecture is based on hybrid Karatsuba multiplication algorithm and the exponentiation module is implemented using Itoh–Tsujii algorithm. The architecture for Itoh–Tsujii algorithm is generally based on field squarer module. However, for FPGA platform, the performance of Itoh–Tsujii algorithm can be improved significantly if it is implemented using field quad module (computing fourth power of the input). Additionally, we have improved the scheduling of the scalar multiplication algorithm by overlapping the mathematical operations of two consecutive scalar bits.

We have validated the proposed architecture for ECC scalar multiplication in $GF(2^{233})$ also. Compared to the existing implementations, our proposed implementation was fastest without having any significant area overhead among reported literature at the time of publishing the work.

8.3.2 Efficient Resource Utilization for ECC Scalar Multiplication in $GF(p)$

In this section, we will focus on the ECC scalar multiplication in $GF(p)$. ECC implementation in $GF(p)$ on FPGA can be significantly improved by deploying FPGA-based hard-IPs like *DSP blocks* and *block RAMs*. In this context, we have proposed a generalized *non-standard tiling methodology* in [15]. The proposed methodology in [15] focuses on the optimum usage of DSP blocks with the objective of faster field multiplications. The multipliers present inside the DSP blocks of modern FPGAs are asymmetric in nature. They are capable of computing 24×17 unsigned multiplication. Due to this asymmetric multipliers, standard school book method of field multiplication will be non-optimum. Hence, we have introduced the nonstandard methodology for field multiplication, which makes the usage of DSP blocks optimum. An example of nonstandard tiling is shown in Fig. 8.5. In [15], we have generalized the notion of nonstandard tiling so that nonstandard tiling can be applied to multipliers with large operand width, which is the typical scenario in case of ECC. An illustration of our proposed methodology is shown in Fig. 8.6.

Using the proposed nonstandard tiling methodology, we have developed a fast field multiplier which uses the DSP blocks in an optimal manner. The application of nonstandard

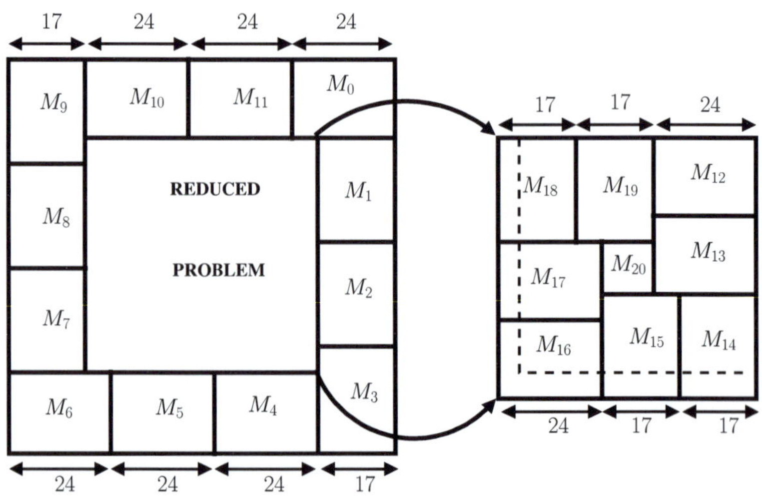

Fig. 8.6 Multiplying operands of width 89 using asymmetric multiplier of dimensions 24 and 17 [15]

Operand width(b)	Mapped Operand width(a)	Decomposition	Reduction Step	Multipliers Required by standard Tiling	Multiplier Required by Non-Standard Tiling
192	191	$24*3+17*7$	$191 \to 47$	96	**90**
224	222	$24*5+17*6$	$222 \to 18$	140	**120**
256	256	$24*5+17*8$	$256 \to 16$	176	**160**
384	382	$24*6+17*14$	$382 \to 94$	368	**360**
521	519	$24*11+17*15$	$519 \to 9$	682	**660**

Fig. 8.7 Nonstandard tiling vs. standard tiling [15]

Table 8.2 Different variant of SBN instruction

Instruction	Memory write-back	Multiplier reset	Key-shift	Right-shift
$SBN_{w\overline{mulks}\overline{rs}}$	✓	x	x	x
$SBN_{nw\overline{mulks}\overline{rs}}$	x	x	x	x
$SBN_{nw\overline{mul}ks\overline{rs}}$	x	x	✓	x
$SBN_{w\overline{mulks}rs}$	✓	x	x	✓
$SBN_{nwmul\overline{ks}\overline{rs}}$	x	✓	x	x

tiling for NIST curves specified for ECC scalar multiplication in $GF(p)$ and the corresponding DSP block requirement is shown in Fig. 8.7.

8.3.3 Lightweight Architecture for ECC Scalar Multiplication in $GF(p)$

In this work [16], we have proposed a single instruction approach for implementation of ECC scalar multiplier suitable for lightweight applications. More specifically, we have built an entire ECC scalar multiplier processor by using only one URISC (Ultimate Reduced Instruction Set Computer) instruction SBN (subtract and branch if negative). It is well known that using such URISC instruction, one can execute any logical or mathematical operation, leading to a Turing complete computer processor. However, as ECC involves many computationally intensive field operations, a stand-alone SBN-processor for ECC scalar multiplier execution will be drastically slow.

To tackle this problem, we have integrated the SBN processor with dedicated hard-IPs (Block RAM, DSP Blocks) of the modern FPGAs to demonstrate an implementation of immensely lightweight, yet practical ECC architecture [16]. We have proposed four different flags in the processor architectures which when set, activates different optimization strategies integrated inside the processor architecture. The optimized strategies proposed by us are: the option of switching off memory write back, dedicated right shifter, and dedicated field multiplier built exclusively with DSP blocks. It must be noted that field multiplication and right shift can also be executed by repeated execution of SBN instruction, but that will be extremely slow and inefficient. Hence, we have developed efficient architectures for these operations using FPGA-based hard-IPs. The details of the proposed SBN instruction along with four different flags are shown in Table 8.2. The architecture of the proposed SBN-processor for ECC scalar multiplication is shown in Fig. 8.8.

For comparing with other existing implementation, we have implemented *NIST P-256* ECC scalar multiplier using SBN processor on Virtex-5 and Spartan-6 platform. In both cases, the slice consumption of the design is less than 100. To the best of our knowledge, this is the first implementation of ECC which requires less than 100 slices on any FPGA device family. The stand-alone SBN processor is itself very lightweight, and the dedicated multiplier core is designed by judicious use of DSPs. Additionally, the block RAMs are used extensively to implement both data and instruction memory of SBN-ECC processor. It must be noted that a designer can choose a budget of slices and block RAMs and can design the corresponding SBN-ECC processor as per his choice. Moreover, the timing performance of the SBN-ECC processor is comparable with the existing implementations. This may seem to be counterintuitive as the proposed SBN processor needs to execute a large number of instructions to complete one single scalar multiplication. However, it must be noted that the proposed SBN processor is coupled with a dedicated field multiplier built using high-performance DSP blocks. This improves the timing performance of the proposed architecture significantly.

8.4 PUFs: Design and Usage in IoT Security

The idea behind PUFs is to utilize device-specific random intrinsic features for identification. Depending on the technology used for implementation, PUFs can be categorized into four categories: Optical PUF, Silicon PUF, Coating PUF,

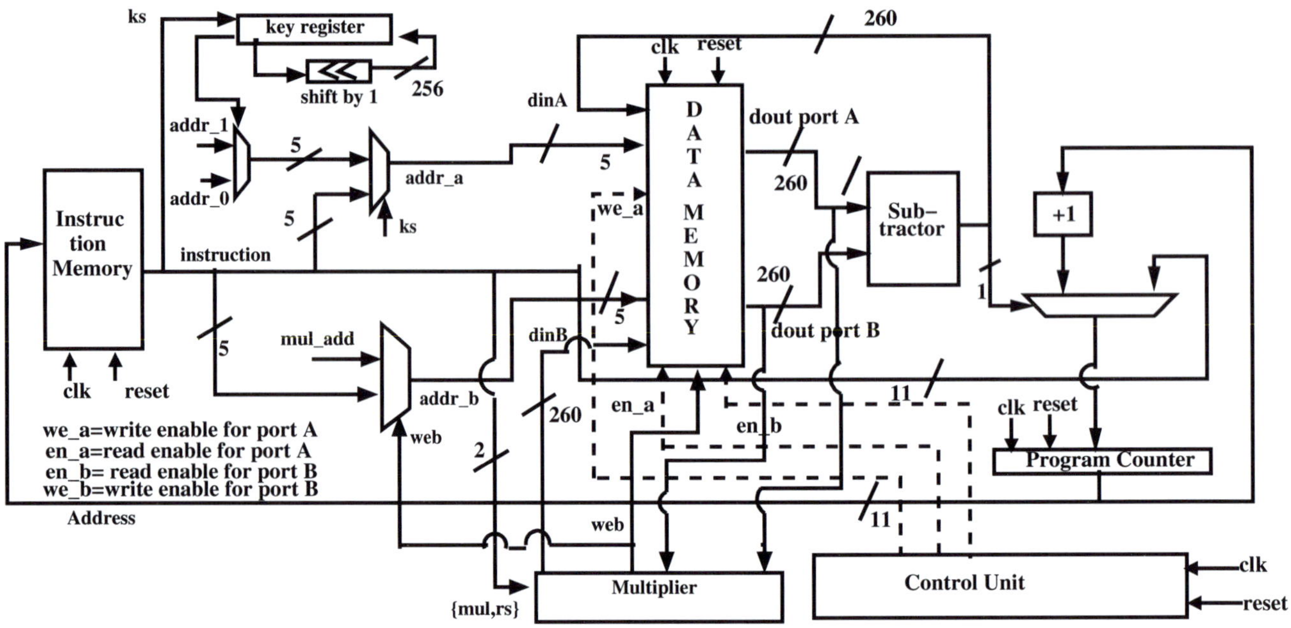

Fig. 8.8 ECC SBN processor architecture [16]

and Acoustic PUF, out of which Silicon PUFs have gained much attention. A silicon PUF is a one-way function embedded in an IC which utilizes the uncontrollable and intrinsic physical characteristics of the IC occurring due to the variations in the manufacturing process. The output of a PUF instance, or *response* is uniquely determined by the input, also known as *challenge* and device-specific variations. An applied challenge and its measured response is generally called a *Challenge–Response Pair* or CRP and the relation enforced between challenges and responses by one particular PUF is referred to as its CRP behavior. Every PUF instance exhibits a unique and unpredictable challenge–response behavior that is hard to characterize physically or mathematically, but is otherwise easily and reliably evaluated, which makes PUF a good candidate for various security applications such as random number generation, key generation, hardware authentication, etc.

Ever since the advent of PUFs, multiple attempts have been made to compromise the security of the hosting device or application by learning the embedded PUF behavior. Machine Learning attacks, being a noninvasive method, have been a very popular choice. In such attack scenarios, a relatively small subset of challenges along with their respective responses is collected by the adversary, attempting to come up with a model describing the challenge–response behavior of the PUF, and are commonly known as classical ML Attacks. An effective countermeasure for this attack is to increase the modeling complexity of the learning algorithm, which in turn would require more training data to create a fairly accurate model. It was later pointed out by Becker in [17] that beside the challenge– response behavior, reliability of a PUF also holds essential information. Reliability of a PUF is a measure to check similarity in response to a particular challenge when applied multiple times to a particular PUF under different environmental conditions. On investigating for Arbiter PUF (APUF), which is an intrinsic delay-based PUF, it is observed that the delay difference ΔD for a specific challenge, which decides the response of APUF is directly proportional to the unreliability of the corresponding response bit if the environmental conditions are kept stable. This is due to the fact that the various sources of noise add an approximately Gaussian delay $D_{noise} = norm(\mu, \sigma)$ to the delay difference ΔD. If the delay difference ΔD_{PUF} for a given challenge C is very large, it is unlikely that the noise term D_{noise} changes the sign of ΔD. Hence the response bit remains the same. However, if the delay difference ΔD_{PUF} is close to zero, then it is much likely to change the response bit. Change in the environmental conditions such as temperature or supply voltage has a similar or rather a stronger effect than thermal noise over the reliability of response. Based on this observation, Becker proposed a reliability-based machine learning attack in [17] using CMA-ES (Covariance Matrix Adaptation-Evolution Strategies) algorithm, which selects a set of models whose reliability vector fits closely to the actual reliability vector of the PUF to be modeled. In each round, certain modifications are made to the selected models and using the fitness criteria, the next generation of PUF models are selected.

Considering these factors, we at SEAL, IIT Kharagpur, have designed a MUX-based Arbiter PUF composition (MPUF) [18], denoted by (n, k)-MPUF, as shown in Fig. 8.9 with the objective of improving modeling robustness and reliability. As shown in Fig. 8.9, there are n data input APUFs and k selection input APUFs and the response of

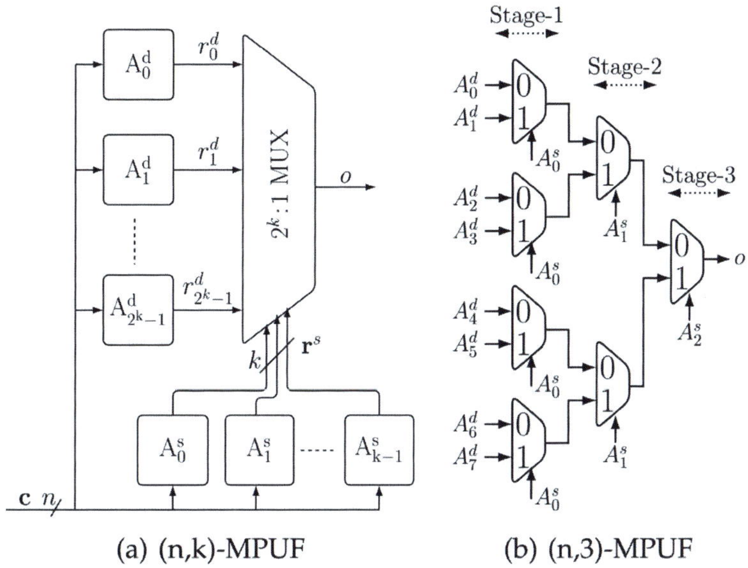

Fig. 8.9 Block diagram of a MUX PUF circuit [18]

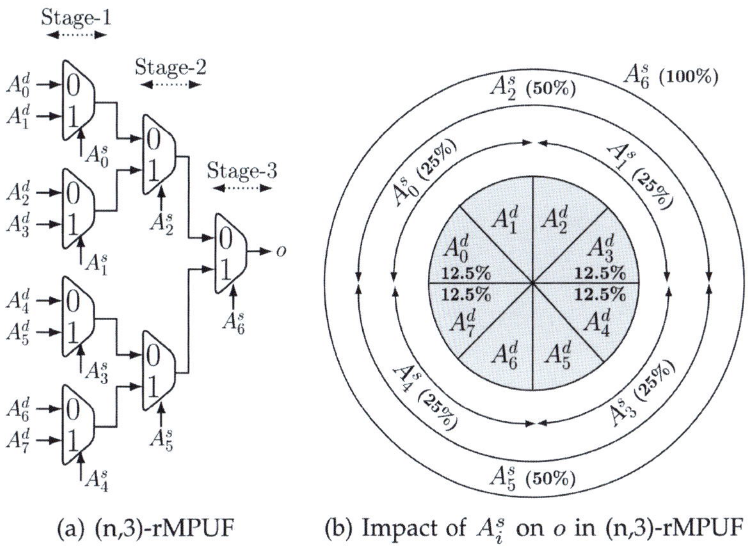

Fig. 8.10 Block diagram of a Robust MUX PUF (rMPUF) circuit [18]

(n, k)-MPUF depends on k selection input APUF and one input data APUF, which is selected by the output of selection APUFs. This implies that the final response is independent of remaining $(n-1)$ APUFs, which results in better reliability. Along with this, a robust MPUF variant, shown in Fig. 8.10, is also proposed which aims to reduce the contribution of selector inputs APUFs on the MPUF CRP space, thereby reducing the correlation between reliability of MPUF and selection APUF outputs, thus making it robust against reliability-based modeling attack. Reduction of contribution in CRP space implies that an adversary would need more CRPs to build a high accuracy model which is required in the training phase of such attacks (Fig. 8.11).

For an (n, k)-rMPUF, the total number of distinct CRPs required for modeling all selection input APUFs is

$$N_k^S = N_k^S + \sum_{j=1}^{k-1} \left(\frac{N_k^S}{2} \times 2^j \right) = 2^{(k-1)} N_k^S \tag{8.2}$$

where N_k^S denotes number of CRPs required for reliability-based modeling of a selection input APUF instance in (n, k)-rMPUF. Thus modeling complexity increases exponentially with respect to number of selection input APUFs k which makes rMPUF robust against Becker's attack.

Moreover, MPUF and its variants have near ideal uniformity and uniqueness values for various values of n, k, and α, as shown in Table 8.3. This work is an attempt to explore the capabilities of MPUF and its variants across various PUF performance metrics and to provide a better alternative to XOR-PUF, which is considered a good choice in PUF-based

Table 8.3 Performance metrics (%) of simulated (n, k)-MPUF/cMPUF/rMPUF

PUF	n	k	$\alpha^{\star}$	Uniformity (Avg., Std.)	Uniqueness (Avg., Std.)	Reliability (Avg., Std.)			$x^{\dagger}$
						(n, k)-MPUF Variant	(k+1)-XOR APUF	x-XOR APUF	
MPUF	64	3	1/2	(50.82, 3.32)	(50.04, 1.43)	(86.24, 0.63)	(77.47, 1.22)	(53.60, 0.50)	11
			1/20	(50.33, 2.72)	(50.02, 0.61)	(98.91, 0.05)	(98.28, 0.10)	(94.62, 0.23)	
			1/80	(50.33, 2.71)	(50.02, 0.61)	(99.74, 0.01)	(99.59, 0.03)	(98.82, 0.05)	
		4	1/2	(49.57, 2.03)	(50.01, 0.61)	(83.57, 0.59)	(71.99, 1.07)	(50.16, 0.09)	20
			1/20	(49.80, 1.67)	(50.01, 0.32)	(98.67, 0.06)	(97.82, 0.11)	(89.40, 0.35)	
			1/80	(49.79, 1.66)	(50.01, 0.32)	(99.69, 0.01)	(99.49, 0.02)	(97.79, 0.06)	
	128	3	1/2	(49.68, 2.08)	(50.02, 0.64)	(86.40, 0.48)	(77.83, 0.91)	(53.74, 0.40)	11
			1/20	(49.79, 1.82)	(50.01, 0.39)	(98.92, 0.04)	(98.31, 0.08)	(94.69, 0.15)	
			1/80	(49.79, 1.81)	(50.01, 0.39)	(99.74, 0.01)	(99.60, 0.02)	(98.84, 0.04)	
		4	1/2	(49.87, 2.24)	(50.00, 0.39)	(83.61, 0.36)	(72.13, 0.74)	(50.15, 0.08)	20
			1/20	(49.95, 1.73)	(50.00, 0.21)	(98.68, 0.04)	(97.85, 0.08)	(89.47, 0.25)	
			1/80	(49.95, 1.72)	(50.00, 0.20)	(99.70, 0.01)	(99.50, 0.02)	(97.80, 0.05)	
cMPUF	64	4	1/2	(49.80, 2.27)	(49.99, 0.23)	(83.76, 0.63)	(71.80, 0.89)	(52.64, 0.43)	12
			1/20	(49.86, 1.77)	(50.00, 0.16)	(98.69, 0.06)	(97.82, 0.08)	(94.09, 0.25)	
			1/80	(49.86, 1.76)	(50.00, 0.16)	(99.69, 0.01)	(99.48, 0.02)	(98.71, 0.05)	
		5	1/2	(50.25, 2.09)	(50.00, 0.18)	(78.82, 0.70)	(64.85, 1.15)	(50.13, 0.08)	21
			1/20	(50.23, 1.64)	(50.00, 0.13)	(98.10, 0.08)	(96.71, 0.17)	(87.09, 0.29)	
			1/80	(50.23, 1.62)	(50.00, 0.13)	(99.55, 0.02)	(99.22, 0.04)	(97.12, 0.08)	
	128	4	1/2	(49.65, 1.83)	(50.00, 0.14)	(83.78, 0.43)	(71.83, 0.98)	(52.66, 0.23)	12
			1/20	(49.75, 1.46)	(50.00, 0.10)	(98.69, 0.04)	(97.82, 0.11)	(94.12, 0.17)	
			1/80	(49.74, 1.45)	(50.00, 0.10)	(99.70, 0.01)	(99.49, 0.03)	(98.72, 0.04)	
		5	1/2	(49.89, 1.34)	(50.00, 0.11)	(78.84, 0.40)	(64.85, 0.64)	(50.12, 0.08)	21
			1/20	(49.90, 1.03)	(50.00, 0.10)	(98.11, 0.05)	(96.73, 0.10)	(87.05, 0.20)	
			1/80	(49.91, 1.02)	(50.00, 0.10)	(99.55, 0.01)	(99.23, 0.03)	(97.11, 0.05)	
rMPUF	64	3	1/2	(50.01, 4.40)	(49.88, 1.02)	(85.10, 0.41)	(78.67, 0.78)	(60.25, 0.45)	15
			1/20	(50.04, 3.80)	(49.95, 0.69)	(98.67, 0.07)	(98.02, 0.11)	(92.93, 0.17)	
			1/80	(50.02, 3.81)	(49.95, 0.69)	(99.68, 0.02)	(99.50, 0.04)	(98.14, 0.08)	
	128	3	1/2	(49.60, 2.24)	(49.98, 0.62)	(85.02, 0.32)	(78.74, 0.57)	(60.28, 0.35)	
			1/20	(49.58, 1.98)	(49.99, 0.41)	(98.66, 0.05)	(98.01, 0.10)	(92.95, 0.14)	
			1/80	(49.59, 1.97)	(49.99, 0.41)	(99.68, 0.02)	(99.50, 0.03)	(98.13, 0.05)	

$^{\dagger}x$ represents the number of APUFs used in the MPUF or rMPUF
$^{\star}\sigma_{\text{noise}} = \alpha\sigma$, where $0 \leq \alpha \leq 1$

authentication protocol. Since rMPUF overcomes the limitations of XOR-PUF, this proves to be a much better alternative to put to practice.

8.4.1 Design of PUF-Based Protocols

One of the major security challenges in IoT framework is the authentication and key management of potentially billions of heterogeneous devices deployed in the network. We try to address this problem and provide a lightweight and secure solution amalgamating the concepts of PUF, Identity-Based Encryption (IBE), and key hash function. Conventional PUF-based protocols are accompanied with the challenge of storing and securing the secret CRP database at the verifier end. In order to offload storage requirement from verifier and to eliminate risk of getting CRP database compromised, we propose a new mechanism wherein we store just a single key in the Non-Volatile Memory of verifier for authentication of all prover nodes under it using a key-ed hash function, thus reducing the space required drastically. This way it would be easier to protect a single key instead of securing a whole CRP database. Additionally instead of using CRP database directly we generate a new security association information between prover and verifier that hides the correlation between the challenge and response of the PUF and can be stored as public information.

We have used this protocol on a video surveillance camera and tested its effect in protecting the device against "man-in-the-middle" and "replay" attacks. Figure 8.12a, b presents a

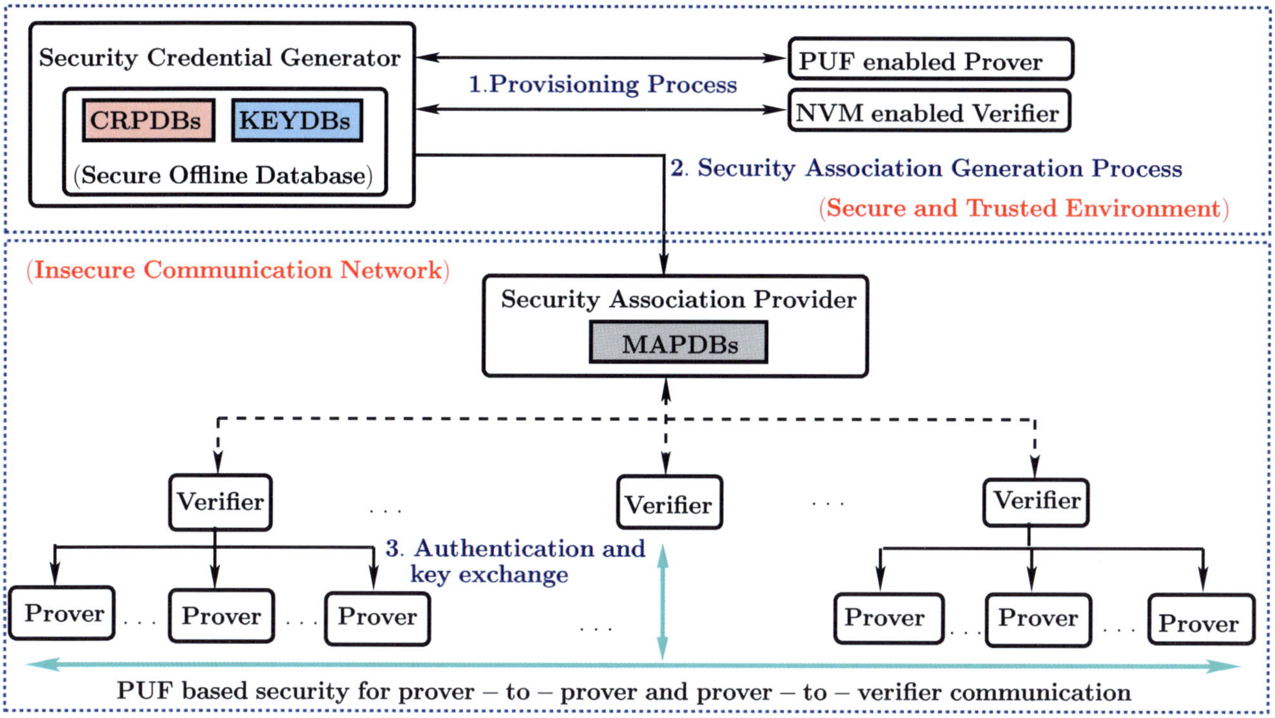

Fig. 8.11 Hierarchical IoT architecture and the proposed secure communication mechanism [19]

practical attack on a video surveillance camera which is successful in the absence of a PUF-based authentication mechanism while Fig. 8.12c, d shows the prevention of the same attack in the presence of our proposed authentication mechanism. The attack was conducted on a Logitech HD UVC camera, which is connected to Intel Edison Board via USB interface to form an IoT node. An mpg-streamer software is run on Edison Board to capture video and send it to the receiver via WiFi, which is then displayed in a web browser using the IP Address of Edison Board. Using hacking tools, the attacker can break the video streaming from the IoT node end and stream pre-captured video from attackers end, hence compromising security. In the presence of our protocol, when an attacker tries to de-authenticate a valid IoT source node and authenticate a malicious node, the receiver will ask for re-authentication, which would require the correct PUF instance, hence leading to a failure. The power consumption of the circuit is reported to be as low as 0.044 W, which makes it suitable for the IoT framework.

8.5 Micro-architectural Attacks and Countermeasures

Micro-architectural features leave a footprint in the processor which is often captured by side channels. In recent microprocessors, various architectural components are incorporated in the system to improve the system performance and these are emerging as new sources of side-channel leakage. Recent micro-architectural attacks named Spectre and Meltdown have taken the world by storm by uncovering two processor vulnerabilities. These vulnerabilities were there in the processor design for decades and uses architectural constructs such as the branch predictors and speculative execution optimization to realize a practical breach of security. We illustrate that the cache memory execution footprints when analyzed with the knowledge of the underlying cache memory structure and characteristics lead to leaking the secret key bytes of block ciphers. On the other hand, when asymmetric-key cryptographic algorithms are implemented and the branch-predictor hardware is monitored, the ciphers are subjected to side-channel attack because of their key-dependent input sequences. We also demonstrate a software-driven fault attack using row-hammer to induce bit flips in the cryptographic secret located in DRAM, by repeated charging and discharging of the memory elements located in the same DRAM bank as the cryptographic secret. The most prominent attack algorithms conceived in SEAL using the micro-architectural primitives are listed below in more details.

8.5.1 Cache Timing Attack on Clefia

We start with the work presented in [20], which discusses the performance of cache timing attacks on Clefia. Clefia has a generalized Feistel structure and unlike other Feistel struc-

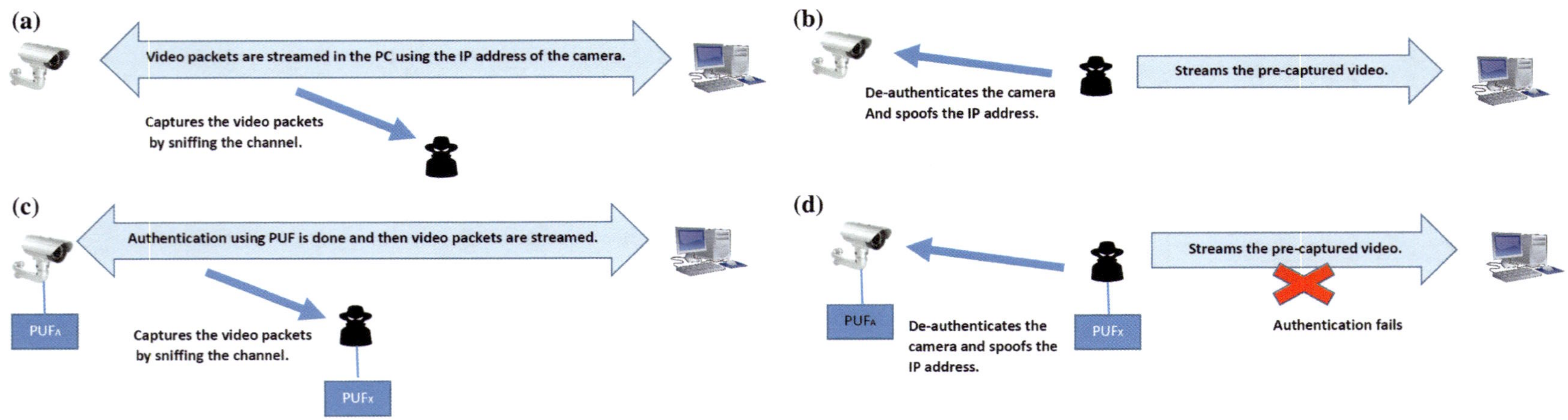

Fig. 8.12 Attack on video surveillance system and protection against it: **a** and **b** show the successful attack in the absence of PUF-based authentication mechanism, while **c** and **d** show the prevention of the attack in the presence of the proposed PUF-based authentication system [19]

tures, it has been implemented using small tables. This was the first attack of its kind to propose a full-scale cache timing attack on ciphers with small tables. The attack uses the fact that parallelization and pipelining of memory accesses can only be performed within a single round of a cipher, and not across rounds.

Among the several design challenges which are supported by today's microprocessors in order to reduce the miss penalty the most important are speculative loading, out-of-order loading, prefetching, parallelization, and overlapping. Speculative loading enables data to be loaded into the cache memory before preceding branching instructions are evaluated and resolved. Prefetching is performed when the hardware prefetcher detects a sequence of memory accesses in a specific order. In out-of-order loading, the processor accesses memory elements in a sequence which is not strictly specified by the program. For a block cipher, speculative loading and hardware prefetching has no effect on the execution time because the key-dependent load operations are random in nature. Additionally cache misses are usually handled out of order. In block ciphers like Clefia, outputs from one round are used in the next, while operations within a round are accessed independent of each other. This implies that memory accesses within a round of Clefia can be performed out of order, but adjacent rounds waits for the previous round to complete and thus follows a sequence.

The attack, demonstrated as follows, modifies Bernstein's cache timing attack [22] with the observation that in the first round the cache timing profile for the ith byte of the key k_i can be affected only by those plaintexts which access the same table because accesses to different tables does not contribute to timing deviations in respective timing profiles. Cache accesses in tables other than the table used by the key byte k_i causes unrequired deviations and thus increases the error in the profile for k_i. This was illustrated in our work to generate more accurate timing profiles. The timing measurement was also improved in compared to the measurement techniques there in the literature. Using the `rdtsc` instruction for timing measurement is known to have errors in measurement due to the out-of-order execution in the pipeline. In our timing measurement, the `cpuid` instruction is invoked before the `rdtsc` to flush the pipeline thus reducing errors [21].

Figure 8.14 shows two accesses to the same s-box table with indices $(in_0 \oplus k_0)$ and $(in_1 \oplus k_1)$, where in_0 and in_1 are the inputs and k_0 and k_1 are the key. Cache hits occur when $(in_1 \oplus k_1)$ fall in the same cache line as $(in_0 \oplus k_0)$, in all other cases cache misses occur. A typical illustration on the timing profiles of the known and unknown keys are provided in Fig. 8.13. The profiled cache timing attack has 3 phases: learning, attack, and analysis phase. During the *learning phase*, the adversary uses a known key (or an exact replica of the attack platform) to build a *timing profile* for each key byte. The profile has on the x-axis, all possible values corresponding to the plaintext byte p_{15}, and on the y-axis the average encryption time corresponding to when p_{15} is fixed at a certain value and the remaining input bytes varied randomly. Note that Fig. 8.13b shows the deviation from average encryption time, instead of the actual encryption time. The timing profile is built with 2^{24} encryptions and called the *template* in this phase of the attack. During the *attack phase*, the adversary builds a similar timing profile, only this time for the unknown key byte. Such a profile is shown in Fig. 8.13a. It can be noted that this profile is very similar to the template in Fig. 8.13b, except for a shift, which occurs due to the *equal*

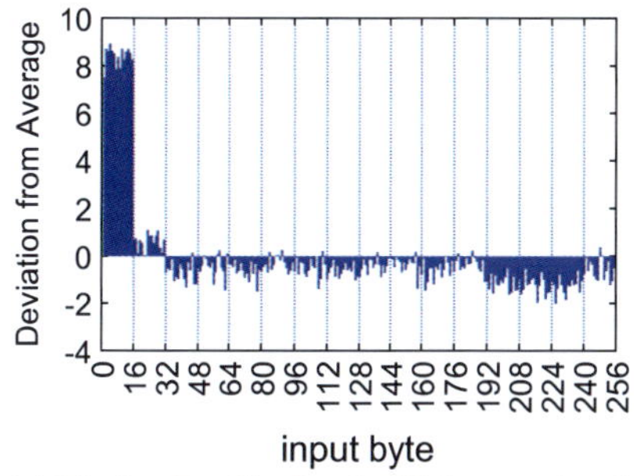

(a) Timing Profile for the Unknown key bytes

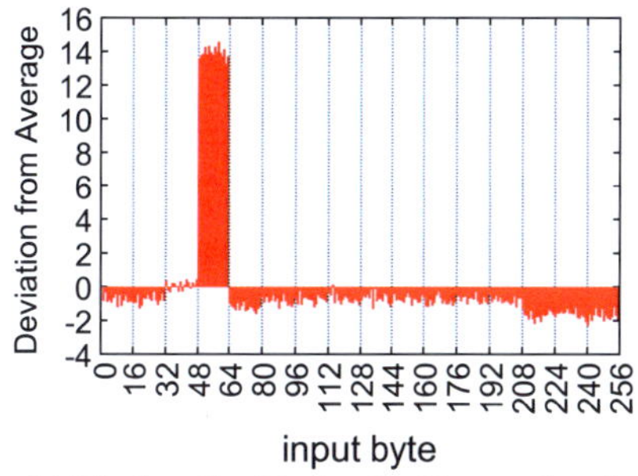

(b) Timing Profile for the Known key bytes

Fig. 8.13 Timing profiles for all possible values of [21]

Fig. 8.14 Simple S-box Lookup structure

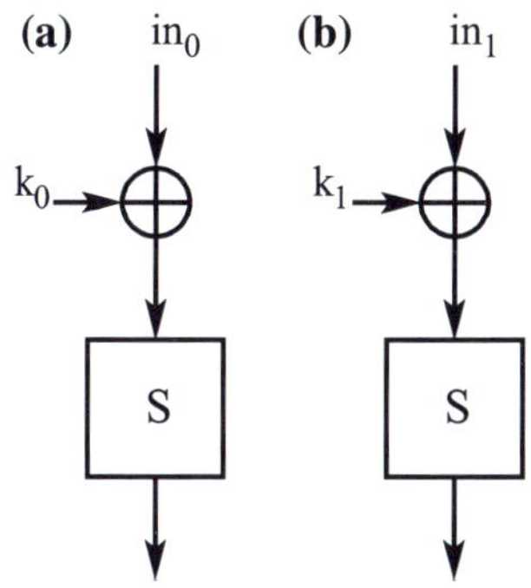

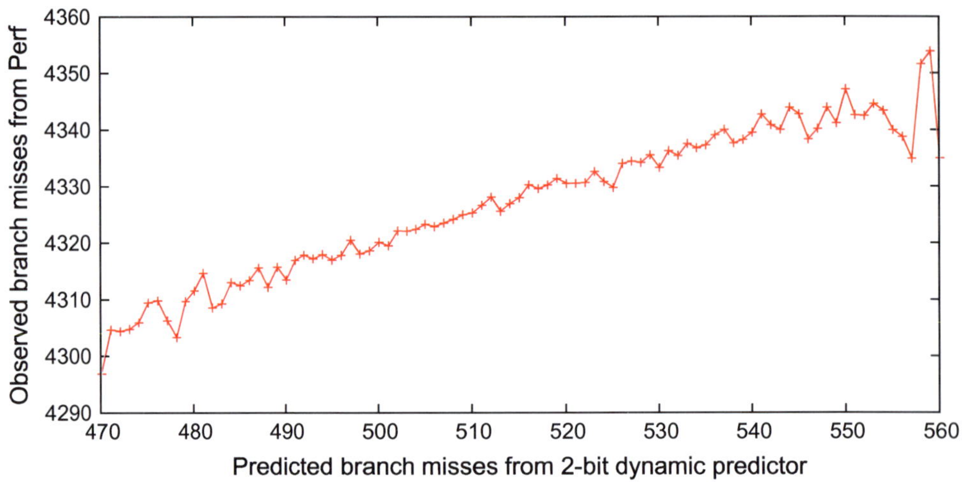

Fig. 8.15 Variation of branch misses from performance counters with increase in branch miss from 2-bit predictor algorithm [23]

images under different sub-keys (EIS) property of the cipher. It is the EIS property that results in the leakage of information about the secret key. During the *analysis phase*, this shift is determined using correlation techniques in order to retrieve the unknown key byte. In a similar manner, timing profiles constructed for other bytes can be used to determine other parts of the key. Our findings show that 121 bits of the 128-bit key can be revealed in $2^{26.64}$ Clefia encryptions on an Intel Core 2 Duo machine.

8.5.2 Branch Misprediction Attack

Asymmetric-key cryptographic algorithms when implemented on systems with branch predictors, are subjected to side-channel attacks exploiting the deterministic branch-predictor behavior due to their key-dependent input sequences. In our work [23], we show that branch predictors can also leak information through the hardware performance monitors. We construct an iterative attack which target the key bits of 1024-bit RSA, where in the offline phase, the system's underlying branch predictor is approximated by a theoretical predictor in the literature. Figure 8.15 illustrates that the 2-bit bimodal predictors bear a direct correlation to the underlying hardware predictor in Intel systems. Subsimulations are performed to classify the message space into distinct partitions based on the event branch misprediction and the target key bit value. In online phase, we ascertain the secret key bit using branch mispredictions obtained from the hardware performance monitors which reflect the information of branch miss due to the underlying predictor hardware. We provide an analysis to justify that the probability of success to guess the ith bit correctly is equivalent to the extent of resemblance of theoretical predictor behavior to the underlying system predictor hardware.

Figure 8.16 shows the correct and incorrect separations for all 4 sets (separated by simulations over two-level adaptive predictor) for the randomly chosen 548th bit location of the target key-stream. Define M_1 as the set which does not cause a miss during multiplication of $(i+1)$th squaring when secret bit $d_i = 1$. Likewise M_2 corresponds to set for a miss. Figure 8.16a plots average branch misses observed from performance counters for each elements in set M_1 and M_2 (each set having L = 1000 elements) and the experiment is repeated over I = 1000 iterations in order to check the consistency of the output. It is evident from the figure that in most of the iterations the average branch miss for set M_2 is more than the branch misses for set M_1 (as expected). We define M_3 and M_4 to be similar sets as M_1 and M_2 which corresponds to $d_i = 0$. Fig. 8.16b plots average branch misses observed from performance counters for each elements in set M_3 and M_4. But we observe an incorrect separation as in most of this case, ciphertexts in set M_4 is having lesser branch misses than in set M_3 which is incorrect since theoretically it should be the reverse. Thus from these two figures, the correct exponent can be easily identified showing correct difference in branch misses.

We also fine-tune this original attack strategy in [24] such that this alternative strategy can retrieve the secret bits much more efficiently and requires much lesser number of inputs. The experimental validations of both the attack strategies are illustrated on public-key cryptographic algorithms as RSA and ECC on several Intel platforms show significant success revealing the secret exponent and scalar value. We also extend our attack to the RSA-OAEP randomized padding procedure where we target the decryption phase of the implementation and the branch miss side-channel information of the entire decoding procedure can be successfully exploited to reveal the secret exponent. What make these results more relevant is the fact that protections which fuzz the timing channels are not sufficient to thwart these attacks, and present performance counters as a distinct side channel which attracts attention to

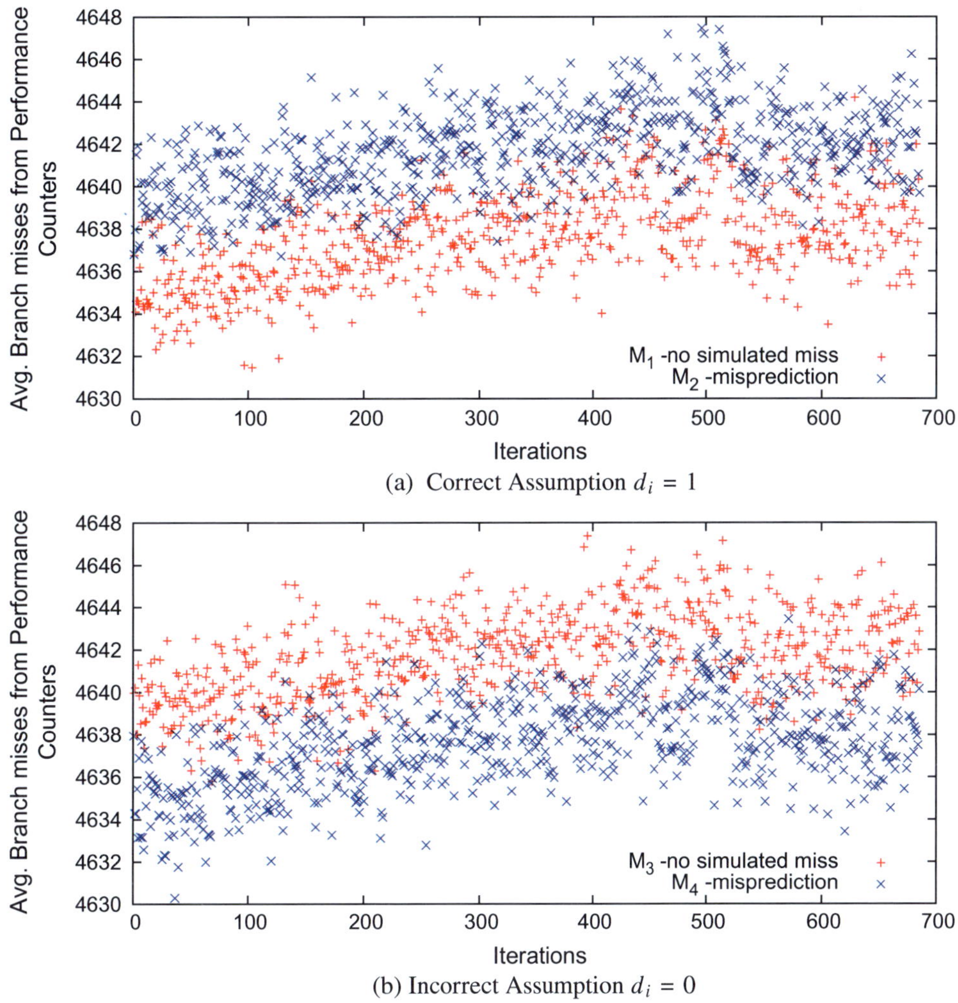

Fig. 8.16 Branch misses from HPCs on square and multiply correctly identifies secret bit $d_i = 1$, ciphertext set partitioned by simulated misses of two-level adaptive predictor [23]

further systematic research to develop systems that are inherently secure.

8.5.3 Software-Driven Fault Attack Using Row-Hammer

Row-hammer is a term coined for a disturbance in DRAM resulting in bit flips due to repeated charging and discharging of DRAM rows in a particular bank. This process is hardware dependent and highly probabilistic but repeatable phenomenon, thus localizing and inducing fault in a specific location of memory is a hard problem till date. Figure 8.17 shows the DRAM architecture and row-hammer is flipping of bits in the DRAM cells neighboring row of heavily accessed DRAM rows.

This work brings into practise a methodology to combine timing analysis to perform row-hammering in a controlled manner to create bit flips in cryptographic keys which are stored in memory [25]. The attack would require only user-level privilege for Linux kernel versions before 4.0 and is unaware of the memory location of the key. An intelligent combination of timing Prime + Probe attack and row-buffer collision is shown to induce bit flip faults in a 1024-bit RSA key on modern processors using realistic number of hammering attempts. This demonstrates the feasibility of fault analysis of ciphers using purely software means on commercial x86 architectures, which to the best of our knowledge has not been reported earlier.

We combine knowledge of reverse engineering of LLC slice and DRAM addressing with timing side channel to determine the bank in which secret resides. The overall idea of the attack involves three steps. The attacker successfully identifies an *eviction set* by following the series of steps as illustrated in Fig. 8.18. The eviction set comprises a set of data elements which map to the same cache set and slice as that of the secret exponent and we identify such set by timing analysis using *Prime* + *Probe* methodology. This set essentially behaves as an alternative to `clflush` statement of x86 instruction set.

The attacker now observes timing measurements to DRAM access by following the series of steps in Fig. 8.19 to determine

(a) A typical DRAM card

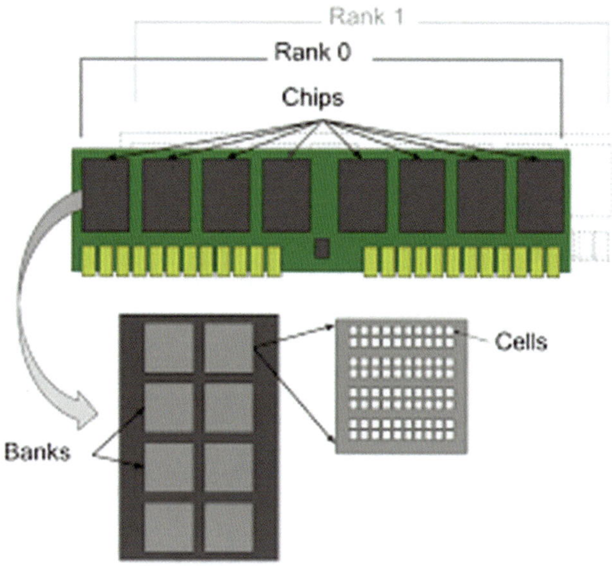

(b) DRAM Architecture

Fig. 8.17 Typical DRAM hierarchy

the DRAM bank in which the secret resides in. The variation in timing is observed due to the row-buffer conflict forced by the adversary.

This leads to repeated opening and closing of rows in DRAM banks inducing bit flips by repeated row activation in the particular bank where the secret is residing. We precisely trigger row-hammer to address in the same bank as the secret. This increases probability of bit flip in the secret exponent and the novelty of our work is that we provide series of steps to improve the controllability of fault induction.

A statistic over observations of bit flips in respective banks is reported in Fig. 8.20. The bar graph shows the number of bit flip instances we were able to observe for respective banks of a single Dual In-line Memory Module (DIMM).

8.5.4 Detection of These Attacks

There are various state-of-the-art countermeasures in the literature to prevent different types of micro-architectural attacks. Some of these countermeasures are implemented by significantly changing the hardware [26], obscuring timing information [27], etc., thereby incurring the cost of extra overhead for their implementations. The severe overhead cost cuts down the performance of the system and increases the energy consumption of the device manifold. A well-established fact is that modern computers will exhibit information leakages, and we need to develop our own defence amid the presence of these leakage avenues, and that too with a low implementation overhead.

Micro-architectural side-channel attacks primarily focus on the execution of the target encryption algorithm and its impact on the behavior of shared hardware resources like CPU-cache, branch-predictor hardware, Dynamic Random Access Memory (DRAM), etc. These attacks on the encryption algorithms manifest the effect of these shared hardware resources in the presence of a concurrently running spy process, which in turn continually monitors the timing or the hardware performance counter statistics of the target secret execution. As a result of these continuous monitoring, the spy processes leave a footprint on the hardware resources. Our objective is to devise a detection module which studies the behavior of these attack algorithms using the low-level hardware events through HPCs. In the presence of these attacks, abnormalities in the number of performance counter events are abrupt in comparison to the normal system execution. In Fig. 8.21, we can observe the distributions of different hardware resources in the presence of various micro-architectural attacks and it is clear from the figure that the micro-architectural attacks work as an anomaly in comparison to the normal behavior of a target system.

Our first approach [28] is to generate a significant volume of low-level data by profiling the hardware performance counters at a granular measurement. We can observe from Fig. 8.21 that attack behaviors are different from the normal process behavior for specific hardware events based on their types. Thus we analyze the cause of abnormality, if any, for the running processes and categorize them into appropriate classes using a pretrained classifier. The result of this phase would help us to comprehend the type of the possible attack process, i.e., whether a cache-based attack or a branch-based attack and so on. We develop a basic template for the target system that we intend to protect against the micro-architectural side-channel attacks, and the process of obtaining information from the system and training a classifier is shown in Fig. 8.22a. However, certifying that a process is actually an attack process and not any high computation process needs further analysis with the performance counter values. The existence of false positives, having an approximately equal

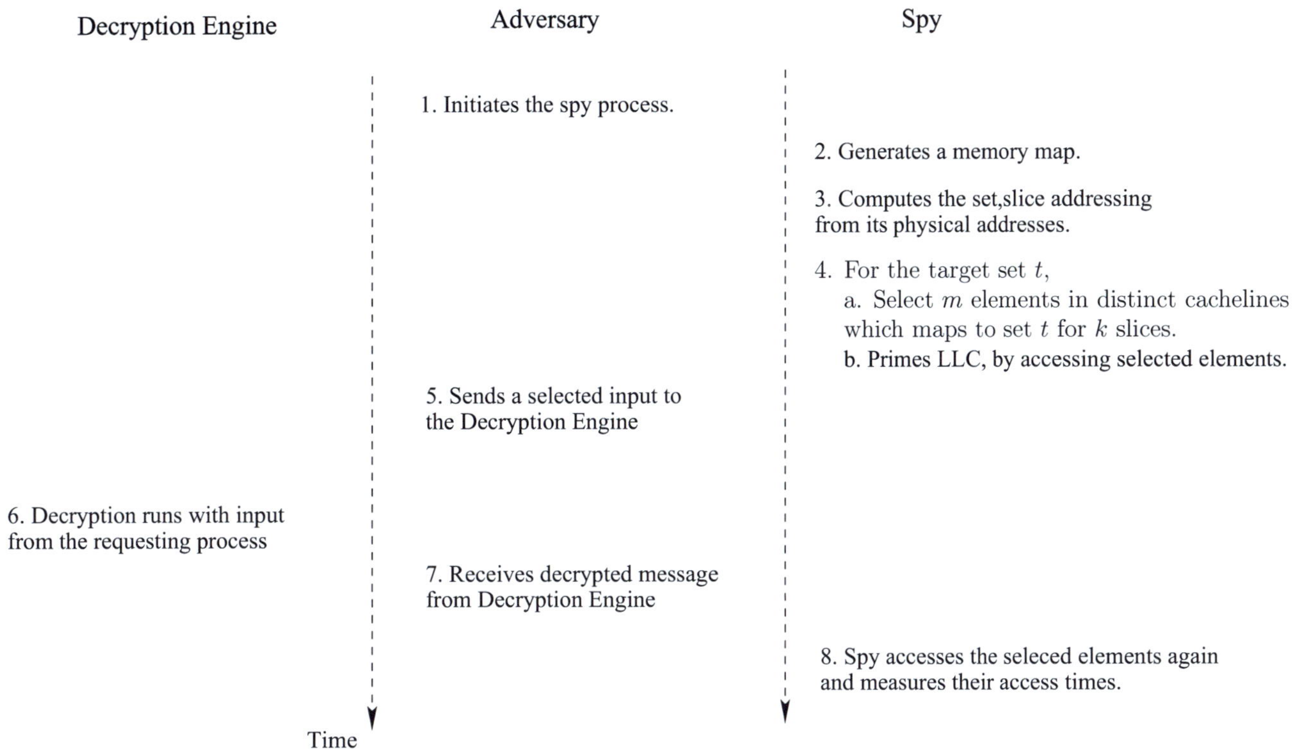

Fig. 8.18 Steps to determine cache sets shared by secret exponent [25]

effect on the performance counter values as that of the targeted attack processes, may confuse the classification phase. Here we introduce the second step of our detection mechanism which removes these false positives by correlating respective hardware events (i.e., cache events for cache-based attacks, branch events for branch-based attacks, etc.) with the secret key of encryption. A true side-channel attack has the highest correlation with the secret key as it needs to repeatedly access the encryption algorithm for retrieving the secret key using statistical evaluations. In order to measure the correlation of the execution trace of an unknown process with the secret key, we collect the HPC values from the target system executing the target encryption algorithm for different plaintexts and a *fixed* secret key. The data collection process is shown in Fig. 8.22b. We store the collected trace for later considering it to measure correlation with an unknown process during the online phase.

In the online phase, we continuously monitor the target system executing the target encryption algorithm and examine whether the system is in a safe state or under the threat of any side-channel attack. The online phase utilizes the classifier and the stored traces prepared during the offline phase. The procedure of the online mode of operation is shown in Fig. 8.23. Based on the category of anomaly obtained from the classifier, the hardware trace of the encryption algorithm is chosen from the stored traces related to the appropriate performance counter events. The correlation value of this trace with the trace obtained by monitoring the anomalous process is then calculated using the Dynamic Time Warping (DTW) method. A low correlation value signifies that the system is in a safe state and the anomalous process is just a high computation program which has no relation to the secret key of encryption, whereas, a high correlation value determines the existence of a possible side-channel attack in the background.

Figure 8.24a shows different nonlinear decision boundaries learned by a Random Forest Classifier from the data obtained by monitoring the hardware events during the offline phase. We can observe that various anomaly models can be well separated from each other based on the acquired data. Figure 8.24b shows the detection of *Cache Timing Attack* on Clefia [29] using the DTW correlation. We can observe from the figure that the attack process has low DTW value with the stored trace in comparison to a benign *Firefox* application, thereby removing the false positives obtained from the classifier.

8.6 Hardware Security to Accelerate Cloud Cryptosystems

Recent progress of cloud-based computation and storage infrastructure have fueled the demand for stronger security for users' data on the cloud. Traditional encryption schemes allow encrypting data to prevent unauthorized access from

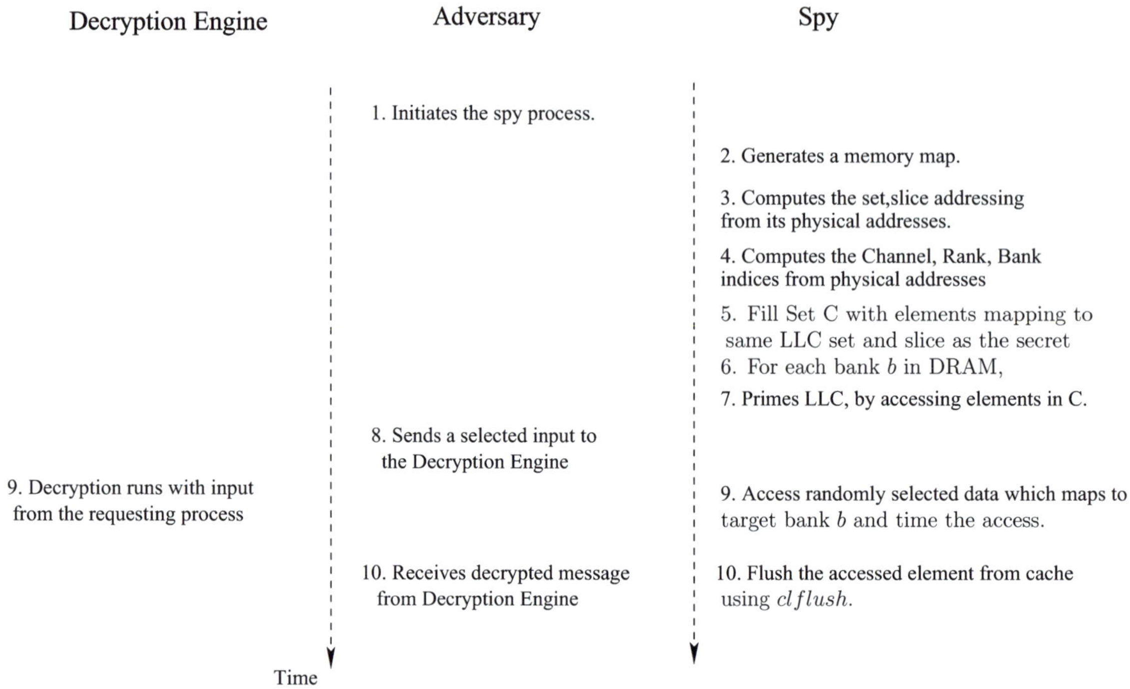

Fig. 8.19 Steps to determine the DRAM bank in which secret maps [25]

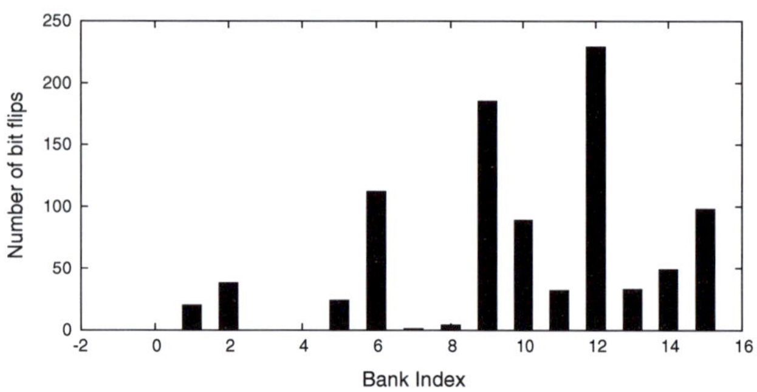

Fig. 8.20 Number of bit flips observed in all banks of a single DIMM [25]

adversaries. However, these schemes fail to provide any extended functionalities to perform particular operations or to evaluate functions over the encrypted data. Searching over encrypted data, performing arithmetic computations over encrypted data, and pattern matching are some of the prominent operations which a cloud infrastructure should provide, but implementing these for an encrypted database, generally in encrypted domain, is a challenging problem in cryptographic research. Currently encrypted domain computation has turned up to be an attractive topic for researchers, development of new generation of functional encryption schemes and homomorphic encryption schemes allow to implement the aforementioned functionalities or operations over encrypted data. Among these, Searchable Symmetric Encryption (SSE) schemes are of special interest, giving the users the ability to search for keyword(s) over encrypted data with restricted access. Traditional SSE schemes employ complex public key based constructs to facilitate the search functionality, and most of those are limited to single keyword search only. This creates a bottleneck for practical implementations preventing deployment at a large scale; primarily for the slower public key based constructions which curtail the throughput of the implementations.

We developed a highly efficient and scalable SSE scheme [30] based on the symmetric key primitives only, with support for multiple conjunctive keyword search. This scheme, referred to as the Hidden Cross Tags (HXT) protocol offers multi-fold improvement over the existing SSE schemes and outperforms other practical implementations which has been validated by software implementation. This scheme incorporates a novel symmetric key based Hidden Vector Encryption (HVE) scheme in the construction. Originally, the Hidden Vector Encryption schemes employ complex Bilinear Map based constructions, which are slow and inefficient, rendering the overall scheme unsuitable for practical use. This Symmetric key Hidden Vector Encryption (SHVE) provides an efficient and more secure symmetric key based approach to reduce the complexity of traditional HVE, thus allowing us to develop a completely symmetric key based SSE. To best of our knowledge, this is the first SSE scheme with complete symmetric key-based construction.

The HXT protocol constitutes of two top-level algorithms, the $SE.EDBSetup()$ and the $SE.EDBSearch()$, with multiple server–client interactions taking place at different

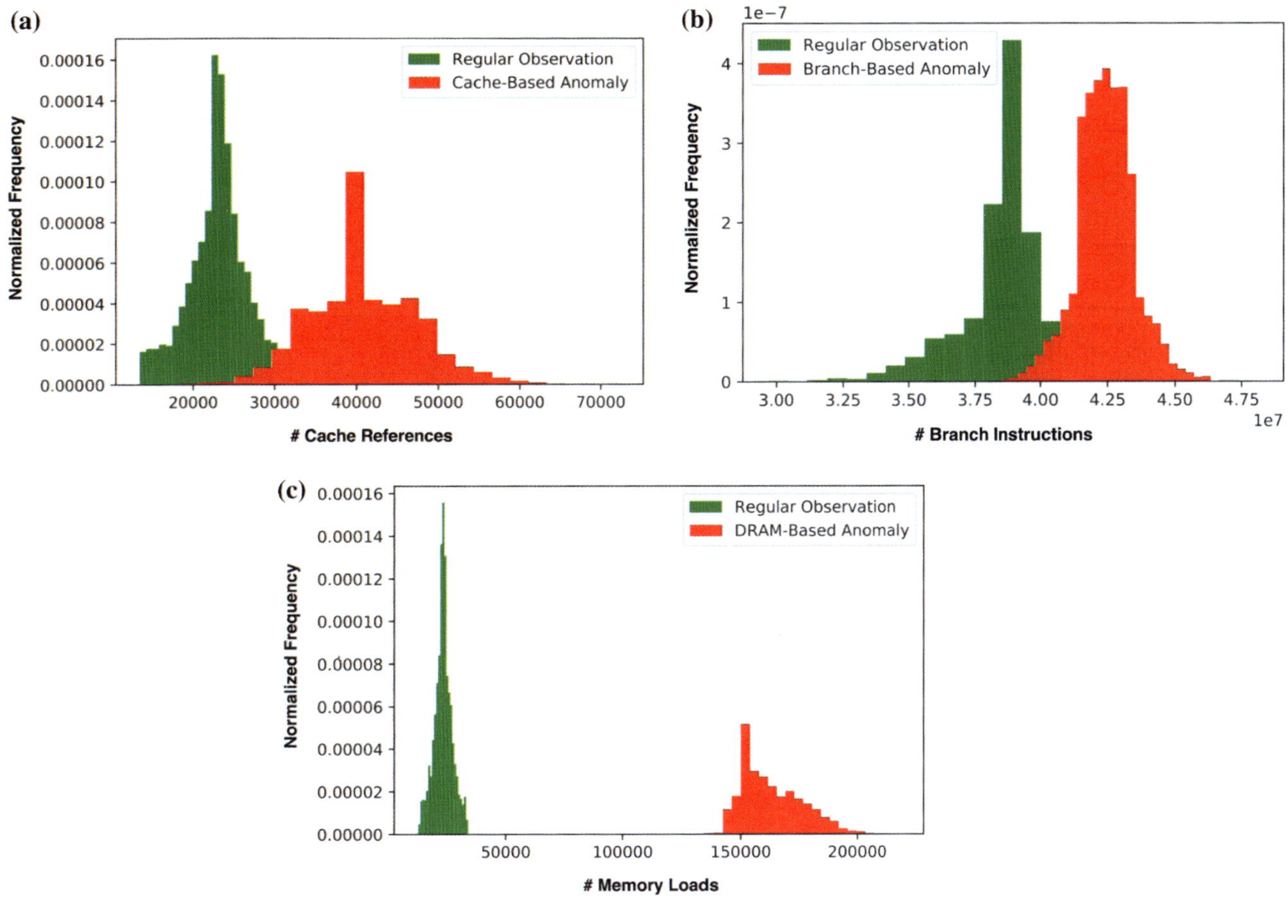

Fig. 8.21 Distinct distributions of **a** Cache-references for regular observations and cache-based anomaly, **b** Branch instructions for regular observations and branch-based anomaly, and **c** Memory-loads for regular observations and DRAM-based anomaly

phases. The $SE.EDBSetup()$ algorithm is responsible for constructing the encrypted database, setting up the system parameters and generating the necessary keys for the protocol. This setup algorithm takes the security parameter λ and the document database DB as input and generates the encrypted database EDB and a master key mk. EDB contains two separate encrypted databases $EDB1$ and $EDB2$, and the master key mk is a collection of several other keys used in the construction. The $SE.EDBSearch()$ algorithm takes the master key mk, query string $q = w_1, w_2, w_3, \ldots, w_n$ and EDB as input and returns the list (encrypted) of document identifiers for the query q.

Additionally, HXT scheme has two other modules, the $TSet$ and the $SHVE$. The algorithms associated with $TSet$ are described below.

- $TSet.Setup(T)$: This algorithm is responsible for the generation of the actual encrypted database $TSet$ from keyword database T, and generation of the corresponding key K_T.
- $TSet.GetTag(K_T, w)$: This algorithm generates a tag $stag$ for a keyword w using the corresponding key K_T.
- $TSet.Retrieve(TSet, stag)$: This algorithm retrieves the document identifier list t corresponding to the keyword w associated with $stag$.

The other major sub-module SHVE has been discussed later in detail. The algorithms belonging to the sub-modules $TSet$ and $SHVE$ are used in different phases of $SE.EDBSetup()$ and $SE.EDBSearch()$. An illustration depicting the server–client interaction at different phases of SSE and the respective algorithms is given in Fig. 8.25. For the complete description of the algorithms $SE.EDBSetup()$ and $SE.EDBSearch()$, the readers may refer to the original paper.

The drastic performance improvement of HXT is primarily due to the SHVE construction. The complete description of SHVE construction is provided below.

Denote a finite set of attributes as Σ and define $\Sigma_* = \Sigma \cup \{*\}$ where $*$ is a wildcard symbol. Define a family of predicates as $\mathrm{P}^{SHVE}: \Sigma \longrightarrow \{0, 1\}$, where for a particular $\mathrm{v} = \{v_1, v_2, v_3, \ldots, v_m\}$, we have a predicate $P_{\mathrm{v}}^{SHVE} \in \mathrm{P}^{SHVE}$ satisfying

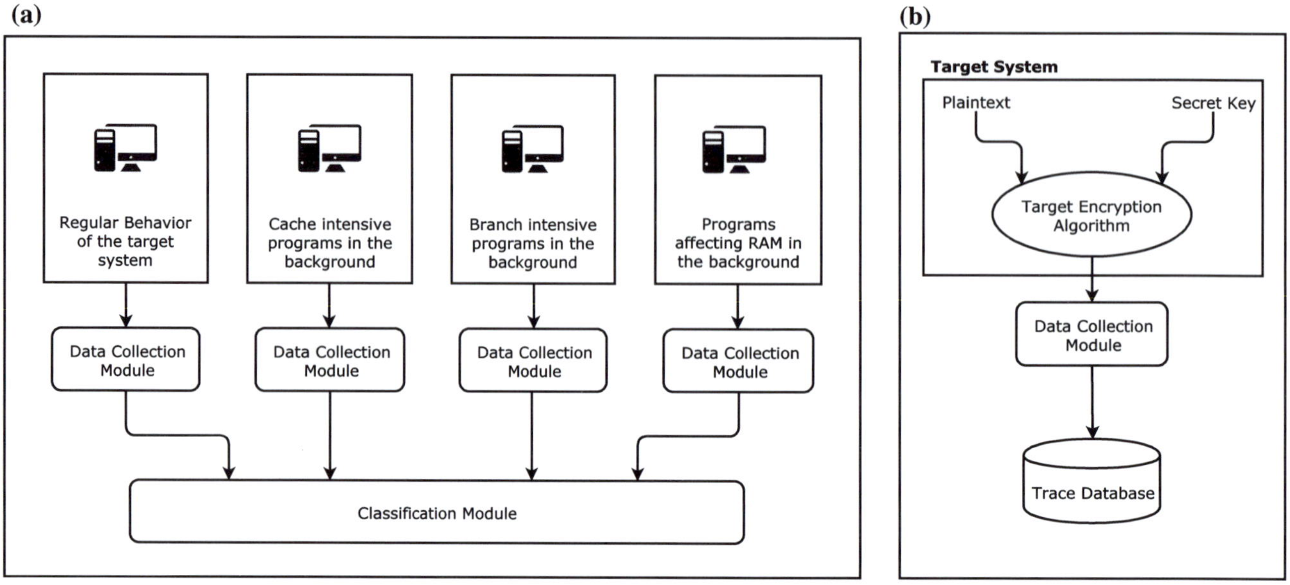

Fig. 8.22 **Offline Phase**: **a** Collection of data for *regular behavior* of the *target system* as well as for different types of *micro-architecture intensive* (like cache-based, branch-based, RAM-based, etc.) programs executed in the background of the target system. **b** Data collection and building a template for a *fixed* secret key considering *target encryption algorithm* (like Clefia, AES, RSA, etc.) for the *target system*

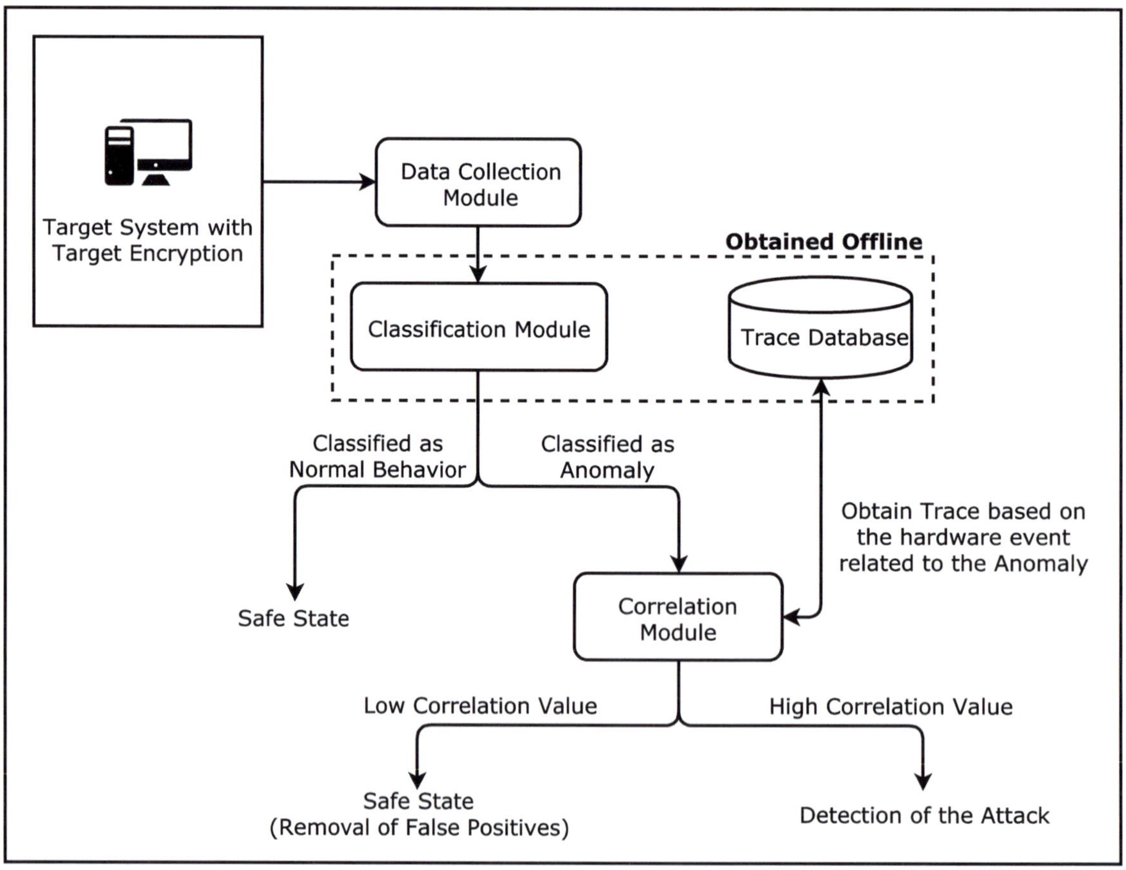

Fig. 8.23 **Online Phase**: Collection of data from the *target system* executing the *target encryption algorithm* and determining whether the system is in *Safe State* or *Under Attack* using the *Classification Module* and *Trace Database* obtained from the *offline* phase

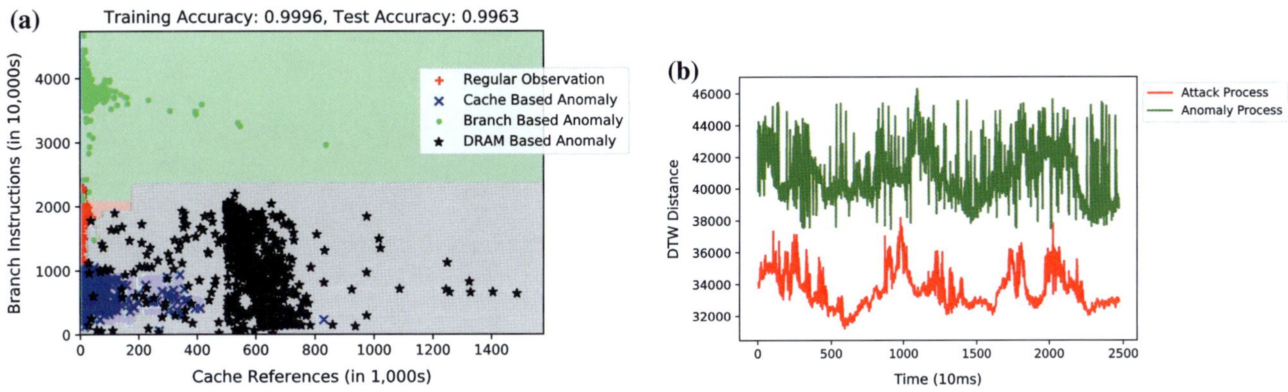

Fig. 8.24 **a** Decision boundaries random forest classifier **b** DTW distances for both cache timing attack on *clefia* (attack process) and *firefox application* (anomaly process)

$$P_{\mathrm{v}}^{SHVE}(\mathrm{x}) = \begin{cases} 1 \; \forall 1 \le i \le m \; (v_i = x_i \text{ or } v_i = *), \\ 0 \text{ otherwise} \end{cases}$$

and $\mathrm{x} = \{x_1, x_2, x_3, \ldots, x_m\} \in \Sigma^m$.

Two cryptographic primitives, Pseudorandom Function (PRF) and a IND-CPA secure symmetric key encryption scheme $Sym.Enc$ are used in this construction. The PRF is defined as the mapping $F_0 : \{0,1\}^{\lambda} \times \{0,1\}^{\lambda+\log\lambda} \rightarrow \{0,1\}^{\lambda+\log\lambda}$ and the symmetric key encryption has both the key space and the plaintext space as $\{0,1\}^{\lambda+\log\lambda}$. The SHVE has the following algorithmic procedures:

- $SHVE.Setup(1^{\lambda})$: Security parameter λ is the input for this algorithm and the algorithm samples the master secret key $msk \longleftarrow \{0,1\}^{\lambda}$. This also defines the payload message space as $M = \{True\}$ and outputs (msk, M)
- $SHVE.KeyGen(msk, \mathrm{v} \in \Sigma_*^m)$: This function takes msk and a predicate vector $\mathrm{v} = \{v_1, v_2, v_3, \ldots, v_m\}$ as input. Denote $S = \{l_j \in [m] | v_{l_j} \neq *\}$ and the locations are $l_1 < l_2 < l_3 \cdots < l_{|S|}$. Randomly sample $k_1, k_2, k_3, \ldots, k_{|S|}$ from $\{0,1\}^{\lambda+\log\lambda}$ and perform the following steps.

$$d_0 = \oplus_{j\in[|S|]}(F_0(msk, v_{l_j} || l_j) \oplus k_j)$$

$$d_1 = Sym.Enc(\oplus_{j\in[|S|]} k_j, 0^{\lambda+\log\lambda})$$

Finally, the algorithm outputs $\mathrm{s} = (d_0, d_1, S)$.
- $SHVE.Enc(msk, \mu =$ 'True', $\mathrm{x} \in \Sigma^m)$: This algorithm takes msk, an input message μ and an index vector $\mathrm{x} = (x_0, x_1, x_2, \ldots, x_m)$ as input and sets the following.

$$c_l = F_0(msk, x_l || l) \text{ for each } l \in [m]$$

next it outputs the ciphertext $\mathrm{c} = (\{c_l\}_{l\in[m]})$

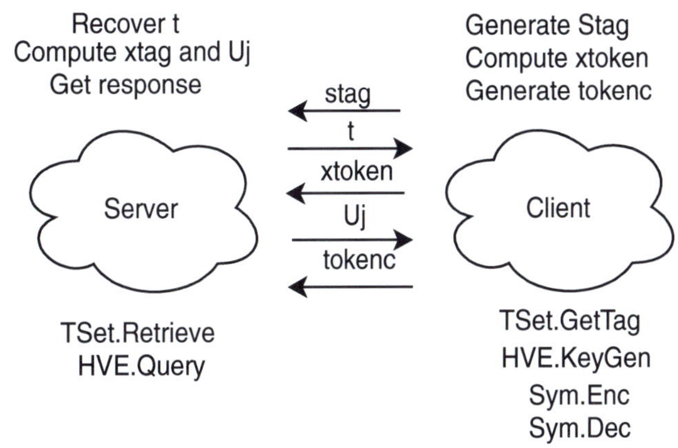

Fig. 8.25 Server–client interaction in HXT

- $SHVE.Query(\mathrm{s}, \mathrm{c})$: Previously computed s and c are passed as input to this algorithm. The algorithm computes

$$k' = (\oplus_{j\in[|S|]} c_l) \oplus d_0$$

and then

$$\mu' = Sym.Dec(k', d_1)$$

If $\mu' = 0^{\lambda+\log\lambda}$, it outputs "True" (the message passed to the input) else it outputs `failure` symbol $\perp$.

The performance of SHVE was evaluated in a software-based prototype implementation. From Table 8.4 it is evident that SHVE outperforms other bilinear map based HVE schemes.

The prototype implementation of HXT was evaluated on a cluster of high-performance workstations. Evaluated with three standard databases from Wikimedia Downloads of size 2.93 GB, 8.92 GB, and 60.2 GB, HXT outperforms other com-

Table 8.4 Performance comparison of SHVE with a bilinear map based HVE (IP) [31]

Scheme	KeyGen(s)	Enc(s)	Query(s)
IP	51.154	50.901	119.219
SHVE	0.172	0.162	0.004

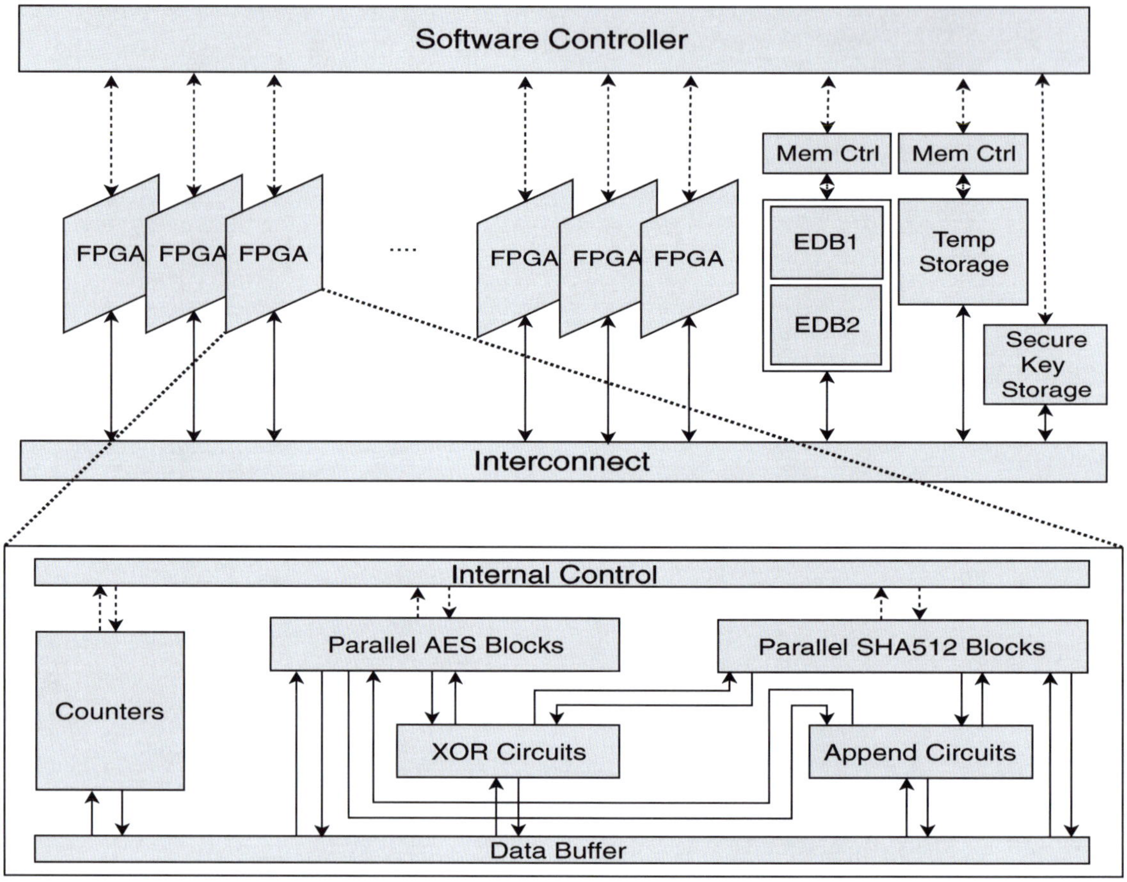

Fig. 8.26 A general architecture for parallel implementation of SSE [31]

petitive schemes by a great margin establishing superiority in this regard.

Software-based implementations provide excellent flexibility in the development of the implementations for SSE, modern-day advanced multicore processors and GPUs can speed-up an architecture for SSE to a great extent. However, a dedicated hardware-based architecture can utilize base-level customization specifically targeted for the SSE scheme. Optimized at basic block level, these hardware-accelerated implementations have the ability to perform exceptionally well compared to the software-based implementations. We proposed a dedicated parallelized hardware-based solution to accelerate the SSE scheme [31]. The HXT protocol incorporates several components which operate independently in a loop. Therefore, multiple scopes for parallelization are present in the architecture. With a meticulously designed parallelized architecture, the throughput of the system can be increased to a great extent. In our solution, we designed the architecture based on a hardware–software co-design environment, where the computation-intensive parallelized operations are accelerated by hardware, and the less computation-intensive operations are moved to the software part. The overall system is controlled by a software-based controller application running on a workstation.

Reconfigurable hardware is an attractive avenue for designers providing accelerated computation for hardware-based implementations. Flexibility, reprogrammability, and abundance of logic resources in modern reconfigurable hardware or FPGAs allow to design, test, and verify a customized design very quickly and provide the users with the ability to update and reevaluate the design within a short period. Evidently, our design is targeted for FPGA-based hardware acceleration. The computational blocks or the cryptographic primitives like AES block cipher, SHA512 hash function are the core part of each of the parallelized instances, as depicted in Fig. 8.26. Interfaced with a high-speed data interconnect and individual node's local controller interface, these modules operate concurrently to speed-up the computation at the base level. Several lightweight designs of the standard cryptographic primitives are available in the literature, capable of delivering high throughput. Multiple instantiations of these blocks do not overload the FPGA, but increases the processing rate to a great extent. Comparing the state-of-the-art software and hardware implementations of the modules, we observe that porting AES to a dedicated hardware gains 31 times speed-up. Similarly, we observe 22 times and 13 times speed-up for SHA512 hash and elliptic curve Curve25519 point multiplication operation on reconfigurable hardware-based design.

Based on this comparison, preliminary analysis indicates that 15 times latency reduction is possible for the preprocessing phase and 20 times latency reduction is possible for the query–response phase in the HXT scheme compared to the parallelized software implementation. Therefore, it is evident that developing a dedicated parallelized hardware accelerator for SSE is crucial for its practical deployment in real-life applications. A general implementation framework for SSE with sound cryptographic techniques, faster implementations with hardware acceleration are essential for SSE and cloud security. This SSE implementation provides a sound and secure design that can be adopted by concerned organizations, and other establishments to build a secure infrastructure for clients, consumers, and citizens of states.

8.7 Conclusions

The article is a summary of the research activities in the topic of Hardware Security performed in our country. We have focused on activities done in the SEAL Lab, Department of Computer Science and Engineering, IIT Kharagpur. The paper deals with several topics in hardware security, starting from hardware design of public-key algorithms, to Side-Channel Analysis in the form of power and micro-architectural attacks. PUFs were discussed as promising primitives for IoT security. Finally, we discussed about lightweight search techniques on encrypted databases, and the potential to develop hardware accelerators for such operations.

Acknowledgements The author would like to thank, in particular, his former and present Masters and Ph.D students at the SEAL Lab, IIT Kharagpur for the various collaborations and findings. In particular, he would like to thank Arnab Bag, Durba Chatterjee, Debapriya Basu Roy, Manaar Alam, Sarani Bhattacharya, Sayandeep Saha, Sikhar Patranabis, Urbi Chatterjee for several discussions during the writing of this article.

References

1. Kocher P, Jaffe J, Jon B Differential power analysis. Advances in cryptology CRYPTO99. Springer, pp 388–397 (1999)
2. Boneh D, Millo R, Lipton R (1997) On the importance of checking cryptographic protocols for faults. Advances in cryptology EUROCRYPT97. Springer, pp 37–51
3. Biham E, Shamir A (1997) Differential fault analysis of secret key cryptosystems. In: B.S.K. Jr. (ed) Advances in cryptology–CRYPTO 1997. Lecture Notes in Computer Science, vol 1294. Springer, pp 513–525
4. Hankerson D, Menezes AJ, Vanstone S (2006) Guide to elliptic curve cryptography. Springer Science & Business Media
5. Christopher, ST (2010) Tarnovsky hacks infineon's 'unhackable' chip, we prepare for false-advertising litigation. www.Engadget.com
6. Sparks ER (2007) A security assessment of trusted platform modules. Technical report, Department of Computer Science Dartmouth College. http://www.cs.dartmouth.edu/~pkilab/sparks/
7. Pappu RS, Ravikanth PS, Recht B, Taylor J, Gershenfeld N (2002) Physical one-way functions. Science 297:2026–2030
8. Tunstall M, Mukhopadhyay D, Ali S (2011) Differential fault analysis of the advanced encryption standard using a single fault. In: Information security theory and practice.security and privacy of mobile devices in wireless communication, Springer pp 224–233
9. Tupsamudre H, Bisht S, Mukhopadhyay D (2014) Destroying fault invariant with randomization. In: Cryptographic hardware and embedded systems–CHES 2014, Springer, pp 93–111
10. Saha S, Mukhopadhyay D, Dasgupta P (2018) ExpFault: an automated framework for exploitable fault characterization in block ciphers. IACR Trans Cryptogr Hardw Embed Syst 2018(2):242–276
11. Saha S, Jap D, Patranabis S, Mukhopadhyay D, Bhasin S, Dasgupta P (2018) Automatic characterization of exploitable faults: A machine learning approach. IEEE Trans Inf Forensics Secur 1 (to appear). https://doi.org/10.1109/TIFS.2018.2868245
12. Güneysu T, Paar C (2008) Ultra high performance ecc over nist primes on commercial fpgas. In: International workshop on cryptographic hardware and embedded systems, Springer, pp 62–78
13. Rebeiro C, Roy SS, Mukhopadhyay D (2012) Pushing the limits of high-speed gf (2 m) elliptic curve scalar multiplication on fpgas. In: International workshop on cryptographic hardware and embedded systems, Springer, pp 494–511
14. Roy SS, Rebeiro C, Mukhopadhyay D (2013) Theoretical modeling of elliptic curve scalar multiplier on lut-based fpgas for area and speed. IEEE Trans Very Large Scale Integr (VLSI) Syst 21(5):901–909
15. Roy DB, Mukhopadhyay D, Izumi M, Takahashi J (2014) Tile before multiplication: an efficient strategy to optimize DSP multiplier for accelerating prime field ecc for nist curves. In: Proceedings of the 51st annual design automation conference, ACM, pp 1–6
16. Roy DB, Das P, Mukhopadhyay D (2015) Ecc on your fingertips: a single instruction approach for lightweight ecc design in gf (p). In: International conference on selected areas in cryptography, Springer, pp 161–177
17. Becker GT (2015) The gap between promise and reality: on the insecurity of xor arbiter pufs. In: International workshop on cryptographic hardware and embedded systems, Springer pp 535–555
18. Sahoo DP, Mukhopadhyay D, Chakraborty RS, Nguyen PH (2018) A multiplexer-based arbiter puf composition with enhanced reliability and security. IEEE Trans Comput 67(3):403–417
19. Chatterjee U, Govindan V, Sadhukhan R, Mukhopadhyay D, Chakraborty RS, Mahata D, Prabhu MM (2018) Building puf based authentication and key exchange protocol for iot without explicit crps in verifier database. IEEE Trans Dependable Secur Comput 1 10.1109/TDSC.2018.2832201
20. Rebeiro C, Mukhopadhyay D (2008) High speed compact elliptic curve cryptoprocessor for fpga platforms. In: International conference on cryptology in India, Springer, pp 376–388

21. Rebeiro C, Mukhopadhyay D, Bhattacharya S (2014) Timing channels in cryptography: a micro-architectural perspective. Springer
22. Bernstein DJ (2005) Cache-timing attacks on aes. Technical report
23. Bhattacharya S, Mukhopadhyay D (2015) Who watches the watchmen?: utilizing performance monitors for compromising keys of RSA on intel platforms. In: CHES. Lecture Notes in Computer Science, vol 9293. Springer, pp 248–266
24. Bhattacharya S, Mukhopadhyay D (2018) Utilizing performance counters for compromising public key ciphers. ACM Trans Priv Secur 21(1) 5:1–5:31
25. Bhattacharya S, Mukhopadhyay D (2016) Curious case of rowhammer: flipping secret exponent bits using timing analysis. In: CHES. Lecture Notes in Computer Science, vol 9813. Springer, pp 602–624
26. Liu F, Lee RB (2014) Random fill cache architecture. In: Proceedings of the 47th annual IEEE/ACM international symposium on microarchitecture, IEEE Computer Society, pp 203–215
27. Martin R, Demme J, Sethumadhavan S (2012) Timewarp: rethinking timekeeping and performance monitoring mechanisms to mitigate side-channel attacks. ACM SIGARCH Comput Arch News 40(3):118–129
28. Alam M, Bhattacharya S, Mukhopadhyay D, Bhattacharya S (2017) Performance counters to rescue: a machine learning based safeguard against micro-architectural side-channel-attacks
29. Rebeiro C, Mukhopadhyay D, Takahashi J, Fukunaga T (2009) Cache timing attacks on clefia. In: International conference on cryptology in India, Springer, pp 104–118
30. Lai S, Patranabis S, Sakzad A, Liu J, Mukhopadhyay D, Steinfeld R, Sun S, Liu D (2018) Result pattern hiding searchable encryption for conjunctive queries. In: Proceedings of the 2018 ACM conference on computer and communications security (To Appear)
31. Bag A, Patranabis S, Tribhuvan L, Mukhopadhyay D (2018) POSTER: hardware acceleration for searchable encryption. In: Proceedings of the 2018 ACM conference on computer and communications security (To Appear)

The World of Bug Bounties—the Indian Scenario

9

Sai Krishna Kothapalli

Abstract

Bug bounties, crowd-sourced security, and responsible disclosures are all different mechanisms that come under the same umbrella. They are avenues an organization or company undertakes to secure themselves by giving ethical hackers the right to report any security vulnerabilities they find in their infrastructure. In case of bug bounties, the incentives ethical hackers receive for finding security vulnerabilities are known as bounties. Though this is a relatively new area, the number of bug bounty hunters has grown significantly in the past few years. This is especially evident in India where the numbers rose to be the highest among all other nations. This article gives an overview of what bug bounties are and how they are utilized by various companies and institutions to secure their online infrastructure. It lays special focus on the Indian scenario and what are the problems that are preventing the Indian organizations from utilizing the full potential of this large talent pool of bug bounty hunters.

Keywords

Bug bounty • Cybersecurity • India • Ethical hacking • Responsible disclosure

9.1 Introduction

In recent times, headlines such as these have become a fairly common occurrence: "Facebook Just Gave An Indian Hacker 10 Lakh Rupees For Finding A Mistake In Their Code!" [1], "This Indian Hacker Has Earned INR 2.2 Crore By Finding Bugs In Facebook, Twitter, And Other websites" [2], etc. On the other hand, headlines such as the following are quite common too: "As Over 100 Govt's Websites Compromised In The Last 12 Months, It's Now The Supreme Court Website" [3], "1.3 lakh Aadhaar numbers leaked from Andhra govt website, linked to personal details" [4], "The official websites of 10 different Indian universities were hacked and defaced on Tuesday" [5], etc.

The present article discusses in detail what exactly are bug bounties. It addresses how Indians are earning such high payments through bug bounty hunting. It also lays out the current cybersecurity scenario in India. In addition, this article discusses how India is leveraging these Indian talents and to what extent.

9.2 What Is a Bug Bounty Program?

A bug bounty program, also called a vulnerability rewards program, is a crowdsourcing initiative that rewards individuals for discovering and reporting software vulnerabilities in a responsible way. These are initiated to supplement internal code audits and penetration tests as part of an organization's vulnerability management strategy. The rewards, which are called bug bounties, are cash rewards ranging from a few to several thousands of dollars. Any hacker from all over the world who finds a bug in the website can report it directly to them and earn money or bounty if their report is valid. The first bug bounty program was launched by Netscape in 1995 [6].

The worldwide cybersecurity landscape has been changing with more complex and frequent attacks. To mitigate this, organizations are not relying only on penetration-testing teams and internal security teams. They are adopting the crowdsourcing model for their cybersecurity. Through bug bounty initiatives, the organizations benefit from a large pool of talented hackers and only pay for the valid vulnerabilities that are reported. They are essentially getting more work done for a reduced investment.

Bug bounty programs have publicly published rules for the program. This includes the do's and don'ts, the scope of the program, the resources they can test, and the resources that are out of scope. Along with these, companies like Dropbox have

S. K. Kothapalli (✉)
Indian Institute of Technology Guwahati, Guwahati, India
e-mail: kothapalli@iitg.ac.in

S. K. Shukla and M. Agrawal (eds.), *Cyber Security in India*, IITK Directions 4,
https://doi.org/10.1007/978-981-15-1675-7_9

Table 9.1 Tabular representation of where hackers are living in the world (after Hackerone Hacker Report, 2018)

Country	Bug bounty hunter (%)
India	23
USA	19.9
Russia	6.3

Table 9.2 Tabular money flow (after Hackerone Hacker Report, 2018)

Countries	Countries where are program located	Where hackers are located
USA	$15,970,63	$4,150,672
Germany	$458,882	$682,528
Russia	$308,346	$1,296,018
UK	$252,960	$916,035
Singapore	$256,280	
Canada	$1,201,485	
UAE	$143,375	
Finland	$142,149	
Malaysia	$138,215	
Switzerland	$118,393	
India	$118,393	$3,098,250
Australia	$118,393	$1,296,411
Hong Kong	$118,393	$749,770
Sweden	$118,393	$746,326
Argentina	$118,393	$673,403
Pakistan	$118,393	$647,339

gone a step ahead and updated their vulnerability disclosure policy which included lines like this—"A pledge to not initiate legal action for security research conducted pursuant to the policy, including good faith, accidental violations."—among many others [7]. These kinds of action from the organizations encourage ethical hackers and bug bounty hunters to search for vulnerabilities with confidence.

Currently, companies like Facebook, Google, Microsoft, etc. are leveraging the power of the crowdsourced security through bug bounties. Facebook paid $880,000 in 2017 [8] and Google paid out almost $3M [9].

9.2.1 Foreign Companies and Bug Bounties

It is thus understood that bug bounties are useful. This is because the organizations can save a lot of money while keeping their online infrastructure secure. This is unlike the conventional ways where it was necessary to pay penetration testers whether they found any vulnerability or not. Now there are more hackers looking at their applications and it is only necessary to pay the ones who submit valid vulnerabilities.

According to a report published by Hackerone, one of the biggest bug bounty-managing platforms in the world, the following conclusions can be drawn (Hacker Report, 2018).

Table 9.1 shows that 23.3% of the bug bounty hunters, which is the highest, are from India followed by 19.9% from USA.

In Table 9.2, the left side consists of countries where the organizations running the bug bounty program are located. The right side depicts where the hackers who got bounties are located. It can be clearly seen from Table 9.2 that the hugest inflow of money is into USA followed by India. But India is nowhere to be seen on the left column, thus highlighting the fact that Indian private companies are not investing any money in bug bounties. This scenario is corroborated by Hackerone (Hacker Report, 2017) as shown in Table 9.3.

According to Facebook Security highlights, 2017—"India came out on top with the number of valid submissions in 2017, with the United States and Trinidad and Tobago in second and third place, respectively [8]."

All this data and many more show that when it comes to cybersecurity, there is an abundance of talent in India. This talent is waiting to be tapped. Looking at the statistics presented here, it can be understood that Indian public sector is not utilizing them.

Table 9.3 Bounties by geography (after Hackerone Hacker Report, 2018)

Country	Where Hackers are earning bounties	Location of company paying bounties
USA	$2,435,169	$6,945,487
India	$1,814,578	$50
Australia	$1,065,095	$24,801
Russia	$723,778	$137,634
Sweden	$633,701	$25,230

Table 9.4 Statistics from Hack the Pentagon program (after [13])

Hacker registered	First report	Bounties paid	Valid reports
1400 +	13 min	$75,000	138

Table 9.5 Statistics of Hack the Pentagon program showing bounties paid (after [13])

Hacker participating	First report	Bounties paid	Valid reports
371 +	5 min	$100,000	118

9.2.2 Indian Private Sector and Bug Bounties

Indian private sector companies are not open to the idea of bug bounties. Though some companies do invest in bug bounties, they pay less bounty which is not at par with the industry standard. This scenario can be seen to be changing slowly in the recent days.

Around May 2017, Zomato faced a data breach [10]. In a blog post on their website, it was explained: "a part of our infrastructure that was used to store user's information was breached by an ethical hacker. The data downloaded because of this breach contained five data points for 17 million users—names, emails, numeric user IDs, usernames, and password hashes. The password hashes leak was a little more contained and impacted a subset of 6.6 million users—all the other users were using Facebook/Google for login. No password information was found for these accounts."

In effect, the personal data of nearly 6 million users was leaked and put up for sale on dark web. This happened because a free webhosting service, 000webhost's user database was leaked in 2015. One of Zomato's developers happened to be using the same username and password for his GitHub account. The mandatory two-factor authentication was not used for all the developer accounts in GitHub at that time. This proved to be a flaw because the Zomato code base is on GitHub.

With the login credentials for the developer, the hacker was able to get into his GitHub account. This allowed the hacker to review one of Zomato's code repositories to which the developer had access. This was not enough to give the hacker access to Zomato's database because it is only accessible to a specific set of IP addresses. The hacker scanned through the code and found a Remote Code Execution vulnerability. Using this, the hacker was able to download the entire database. When contacted by the Zomato team the hacker told that the reason behind this hack is that Zomato had a bug bounty program in which the security researchers are not being rewarded and he wanted them to improve it. Following this, Zomato started rewarding the bug bounty hunters monetarily. Till date, they have paid more than $100,000 in bounties [11].

9.3 Foreign Government

It is old news that governments have not had good relationships with hackers and security researchers in the past. In many cases, even tampering with the systems in any way is a serious offence [12]. However, the times have changed now.

The United States Pentagon started its bug bounty program which was designed to identify and resolve security vulnerabilities within Defense Department public-facing websites through crowdsourced security. The pilot ran from April 18, 2016 until May 12, 2016. It is called "Hack the Pentagon" [13] (Table 9.4).

After the pilot program ended, the United States Department of Defense (US DoD) noted that the vulnerabilities continued being submitted. So, they announced an open-ended Vulnerable Disclosure Policy that did not offer rewards but would legally allow people to submit security vulnerabilities any time. These vulnerabilities should be related to public-facing websites and web applications owned by the US DoD.

And then U.S. Department of Defence started "Hack the Army" (November–December 2016) which paid $100,000 in bounties as shown in Table 9.5.

The "Hack the Air force" program (May–June 2017 and December 2017–January 2018) paid $233,883 in bounties, as can be seen in Table 9.6.

Table 9.6 Statistics of the "Hack the Airforce" program (after [13])

Hacker participating with 30 from outside U.S.	First report	Bounties paid ($130.000 + $103.883)	Valid reports (207 + 106)
275 +	1 min	$233,883	313

Table 9.7 Website defacement statistics of 2016

Number of domains hacked	Top level domains
449	*.ac.in
284	*.gov.in
14	*.nic.in

Some more government agencies around the world like Singapore Ministry of Defense (MINDEF), etc. started to adopt the concept of crowdsourced model of security like bug bounties.

9.4 Indian Government

Now that the fact has been established earlier that India has a large pool of good security researchers, it should be looked into at how good they are being used.

Computer Emergency and Response Team (CERT-IN) is the nodal agency for responding to computer security incidents as and when they occur. National Critical Information Infrastructure Protection Centre (NCIIPC) is designated as the National Nodal Agency in respect of Critical Information Infrastructure Protection. Both CERT-IN and NCIIPC are the DeFacto bodies formally capable of taking care of handling responsible disclosure in India and have some kind of vulnerability reporting mechanism but none of them advertise a proper responsible disclosure policy and what is going to happen after reporting a vulnerability to them and how much legal protection the researcher who reported has, etc. Table 9.7 shows the website defacement statistics taken from CERT-IN [14].

Particularly in October 2016, 222 *.ac.in, 143 *.gov.in, 6 *.nic.in websites were hacked. This hints that a cyber war took place in that month since almost half of the entire hackings happened during that time.

Looking at the 2017 website defacement data as shown in Table 9.8 (except month September), we can see that the numbers remain almost the same. In the *nic.in domain, regular audits take place and hence the rules are somewhat strict. Whereas in the *gov.in domain, the projects are outsourced, and it is up to the third parties to take care of the auditing. In the *ac.in domain, some club websites are developed by students and some conference websites are developed by some third-party contractors. Most of these websites are hosted on out-of-date frameworks. This renders them hugely vulnerable to attacks as can be seen from Table 9.2. The 2018 statistics are not up to date on their website.

The following are two case studies pertaining to security incidents that happened in the recent past of India.

9.4.1 Case Study 1

The author personally found and reported a serious security vulnerability on February 20, 2016 in BSNL [14]. It is a telecommunication company owned by the Indian government. The vulnerability that was found is called a Blind SQL injection. It is very common and well known to researchers in the security community. The vulnerability found was so simple and the fact that it was found on the login page of the intranet site raises some concerns on the quality of the security auditing that is being done.

Various attempts were made to contact BSNL, starting from sending an email to BSNL Chairman and Managing Director with a subject "Security vulnerability in one of the BSNL's website. (Important !!)" There was no response to either of the mails. A week later the author sent another mail to some of other high-level engineers to face similar disinterest. Attempts were made to message BSNL on social media platforms like Twitter and Facebook. Even the Prime Minister of India was tweeted to. But no action was taken. This entire ordeal took over 2 months and as a last resort, the author sent an email to his professor in Indian Institute of Technology, Guwahati. The mentor's advice was to report it to CERT-IN.

Due to various reasons, the author could not report the vulnerability to CERT-IN. Two years later, an anonymous French cybersecurity researcher who goes with the Twitter handle @fs0c131y disclosed the same vulnerability. This hacker, who had nearly 70,000 followers on Twitter at that time, made a public announcement of the vulnerability on Twitter. BSNL patched the vulnerability in 1 day, whereas the same report had been ignored when reported by an Indian researcher 2 years back.

Table 9.8 Website defacement statistics of 2017

Number of domains hacked	Top level domains
196	*.ac.in
136	*.gov.in
14	*.nic.in

Table 9.9 Timeline of events as disclosed by Nemo (after [15])

30 May 2018	Reminder sent to NCIIPC and CERT asking for updates.
19 Apr 2018	Reminder sent to UIDAI asking for acknowledgement.
19 Mar 2018	Confirmation from NCIIPC thanking me for the report..
18 Mar 2018	Confirmation from NCIIPC asking for more details. I replied with a quote of previous exchange.
17 Mar 2018	Notified NCIIPC rvdp@nciipc.gov.in.
15 Mar 2018	Reminder sent. No response.
19 Feb 2018	Acknowledgement from CERT.
19 Feb 2018	Reminder sent to ceo@uidai.gov.in and info@cert-in.org.in.
21 Jan 2018	Reported to ceo@uidai.gov.in and info@cert-in.org.in. No response.
16 Jan 2017	Initially reported to help@uidai.gov.in. No response).

9.4.2 Case Study 2

On September 15, 2018 an Indian Security Researcher named Abhay Rana also known as "Nemo" disclosed this issue [15]. Rana found out that the Unique Identification Authority of India (UIDAI) portal is using an old and discontinued version of a software which resulted in XSS (cross-site scripting) using which any malicious attacker can spoof the content on the page and may even users in to giving away some personal information.

The timeline of events that happened is presented in Table 9.9 (After [15]).

Even after a lot of deliberation the issue was not fixed. This incident shows a similar trend when it comes to the Indian Government.

9.5 Conclusion

In this article, the current scenario of public and government agencies of India has been investigated. In addition, the cybersecurity scene in other parts of the world has been laid out with respect to their ideologies toward hackers and bug bounty hunters. A special emphasis has been laid in detailing how these security researchers are being utilized to protect the online infrastructure of government and private companies alike. In conclusion, why India should utilize the talented Indian hackers and how Indians can benefit from the hackers at home should be studied. Following are the author's suggestions based on the studies conducted:

- The government should implement a proper responsible disclosure framework. The security researchers should be given the right to report the security vulnerabilities they discover in a responsible and legal manner.
- The government should properly coordinate with the security researchers and enquire more details if necessary. The researchers should be updated once the issue has been fixed. Their help can be used again in checking if the fix works.
- The third thing would be to have a good security team inside the government. More cybersecurity professionals should be recruited who are encouraged to be proactive instead of reactive.
- The nodal agencies like CERT-IN and NCIIPC should proactively take steps to inform the software developers in government sector about zero-day vulnerabilities in all kinds of frameworks (like WordPress, Drupal, etc.) which are used to build websites.

- Steps should be taken to educate the developers about the best security practices in software development.
- Tie-ups with private cybersecurity companies should be encouraged.

The Information Technology Act, 2000 of India with a major amendment done in 2008, India as a country has moved forward in terms of technology usage and this law needs an upgrade which is more suitable to the current scenario. It should be updated in a way to protect the security researchers from any kind of harassment and legal threats from the system if they are disclosing any kind of security vulnerability in a responsible way.

In the present-day scenario, almost everything is digitized—from transactions and online shopping to fee payment and registrations for government-sponsored workshops. There is a saying that the next World War is not going to be fought with guns and bombs, rather it is going to be a cyber war. There have been instances where countries have tried to hinder the development of other countries by launching a dedicated cyberattack. In 2016, Russia allegedly meddled with the United States presidential elections to ensure their favored candidate wins [16]. In another instance, the US and Israel combinedly launched a malware attack called Stuxnet on Iran's nuclear program [17]. Additionally, in 2015, Russian hackers successfully compromised information systems of major energy distribution companies in Ukraine. This led to the temporary disruption of electricity supply to the end consumers [18].

In such a hostile environment, India must proactively take steps to protect its critical infrastructure like electricity, water, etc. from state-sponsored cyberattacks.

References

1. Facebook just gave an Indian hacker 10 lakh rupees for finding a mistake in their code! https://www.indiatimes.com/news/india/facebook-just-gave-an-indian-hacker-10-lakh-rupees-for-finding-a-mistake-in-their-code-261992.html
2. This Indian hacker has earned inr 2.2 crore by finding bugs in facebook twitter and other websites 2017 March 07. https://www.huffingtonpost.in/2017/03/07/interview-this-indian-hacker-has-earned-2-2-crore-by-finding-b_a_21874926/
3. Hacked: As over 100 govt's websites compromised in the last 12 months, it's now the supreme court website (2018). https://inc42.com/buzz/hacked-after-114-governments-portals-in-the-last-12-months-its-now-the-website-of-supreme-court-of-india/
4. Around 1.3 lakh aadhaar numbers leaked from Andhra govt website, linked to personal details (2018). https://www.thenewsminute.com/article/13-lakh-aadhaar-numbers-leaked-andhra-govt-website-linked-personal-details-80178
5. A Pakistani group hacked into 10 Indian university websites as revenge against Indian hackers (2017). https://www.indiatimes.com/technology/news/a-pakistani-group-hacked-defaced-10-indian-university-websites-as-retaliation-against-indian-hackers-276456.html
6. Protecting security researchers (2018). https://blogs.dropbox.com/tech/2018/03/protecting-security-researchers/
7. Bug bounty program. https://en.wikipedia.org/wiki/Bug_bounty_program
8. Highlights of year 2017: $880,000 paid to researchers (2018). https://www.facebook.com/notes/facebook-bug-bounty/2017-highlights-880000-paid-to-researchers/1918340204846863/
9. Google's bug bounty programs paid out almost $3m in 2017 (2018). https://techcrunch.com/2018/02/07/googles-bug-bounty-programs-paid-out-almost-3m-in-2017/
10. Security update - what really happened? And what next? (2017). https://www.zomato.com/blog/security-update-what-really-happened-and-what
11. Zomato's bug bounty program paid more than $100,000 (2018). https://hackerone.com/zomato
12. Aaron swartz. https://en.wikipedia.org/wiki/Aaron_Swartz
13. Distributed defense: how governments deploy hacker-powered security (2018). https://www.hackerone.com/sites/default/files/2018-03/Distributed%20Defense-How%20Governments%20Deploy%20Hacker-Powered%20Security.pdf
14. Defacement statistics. https://www.cert-in.org.in
15. The full story of the 2018 BSNL India hack (2018). https://medium.com/@kmskrishna/the-full-story-of-the-2018-bsnl-india-hack-85c98e3f10f8
16. How Russia helped swing the election for trump (2018). https://www.newyorker.com/magazine/2018/10/01/how-russia-helped-to-swing-the-election-for-trump
17. Stuxnet. https://en.wikipedia.org/wiki/Stuxnet
18. December 2015 Ukraine power grid cyberattack. https://en.wikipedia.org/wiki/December_2015_Ukraine_power_grid_cyberattack

10 Post-quantum Cryptography: An Introduction

Shweta Agrawal

Abstract

We present a brief introduction to post-quantum cryptography. This note introduces the concept of post-quantum cryptography, discusses its importance and provides a short overview of the mathematical techniques that are currently used to develop this field.

Keywords

Post-quantum cryptography • Lattices

10.1 Introduction

Cryptography is a rich and elegant field of study that has enjoyed enormous success over the last few decades. At a very high level, cryptography is the science of designing methods to achieve certain secrecy goals, for instance, that of hiding information, so that learning the message from a cryptographically sealed envelope implies a solution to some well-known mathematical problem. By suitably choosing the underlying mathematical problems to be intractable, we may rest assured that an attacker's chances of learning secret information are extremely small; in particular, she must outperform all the mathematical minds that have attempted without success to solve the underlying problem in order to learn the secret. Choosing the underlying hard problem is thus of paramount importance, and we would like to have strong evidence that current day computing resources do not permit an attacker to solve the problem in any reasonable time.

Here, the terms "current day computing resources" and "reasonable time" warrant further investigation. What is considered as reasonable time depends on the application; for securing credit card transactions, we may expect that an attacker will not spend 10 years to break secrecy, but this may not be reasonable for highly sensitive defence communications. The question of computing resources is even more delicate: does the adversary have access to a mobile phone, a laptop, a cluster of computers, or a supercomputer? While again the answer to this question depends on the application, the subject of this note is the very model of computation. Traditionally, cryptography has been based on problems that are conjectured to be infeasible in the realm of classical computers. However, recent times have seen significant advances in the design and construction of *quantum computers*, which are more powerful than classical computers. If an attacker has access to a quantum computer, are known cryptosystems safe?

Two of the most popular problems underlying most current day cryptography are the integer factorization problem and the discrete logarithm problem, please see [1] for a discussion. While the best known classical algorithms to solve these problems take exponential time, a breakthrough work by Shor [2] demonstrated that they can be solved in *quantum* polynomial time. Thus, in the realm of quantum computers, most current day cryptography breaks down. It is therefore necessary to base cryptography of the future on problems that remain intractable against quantum computers. While it is unclear when quantum computers will become a reality, recent times have seen significant strides in this area and it is widely accepted that developing cryptosystems that are secure against quantum computers is an urgent need. To address this, the "National Institute of Standards and Technology" (NIST), a unit of the U.S. commerce department, initiated a process to "solicit, evaluate, and standardize one or more quantum-resistant public-key cryptographic algorithms" [3].

In this note, we do not discuss the progress made in constructing quantum computers, nor the differences between classical and quantum computing. Instead, we study some problems that are conjectured to be quantum hard and discuss some applications to cryptography.

S. Agrawal (✉)
Indian Institute of Technology Madras, Chennai, India
e-mail: shweta@iitm.ac.in

S. K. Shukla and M. Agrawal (eds.), *Cyber Security in India*, IITK Directions 4,
https://doi.org/10.1007/978-981-15-1675-7_10

10.2 Directions for Post-quantum Cryptography

At a high level, the mathematical problems underlying post-quantum cryptography may be categorized into the following broad families:

Lattice-Based Cryptography. Of all known candidates for post-quantum cryptography, perhaps the most popular is lattice-based cryptography. Informally, a lattice is a set of points in an n-dimensional space with a periodic structure. Lattices occur everywhere, from crystals to stacks of fruit to ancient Islamic art, and have been widely studied, starting with ancient mathematicians such as Lagrange, Minkowski, and Gauss up to modern computer scientists. A lattice may be represented using a basis that generates its points, and given a basis, the most basic question that may be posed is that of finding the smallest nonzero point in the corresponding lattice. This classic problem is known as the shortest vector problem (or SVP) and is related to many other lattice problems as we shall see subsequently.

Despite substantial research effort, no efficient quantum algorithms are known for lattice problems that outperform classical ones significantly. In fact, the only advantage quantum computers offer in this regard is modest generic speedups. Besides, lattice-based cryptography has many other advantages. Cryptosystems based on lattices are often algorithmically simple, efficient, and highly paralellizable. Moreover, lattice-based cryptography enjoys a surprising connection between average-case and worst-case hardness [4] which makes it especially attractive. In more detail, cryptography is based on average-case intractable problems, which means that randomly chosen instances of problem must be difficult to solve. On the other hand, complexity theory usually studies worst-case hardness, where a problem is considered hard if there merely exists an intractable instance of the problem. In a surprising work, Ajtai [4] showed that certain lattice problems are hard on the average if some related lattice problems are hard in the worst case. This allows for the design of cryptographic schemes that are infeasible to break unless *all* instances of certain lattice problems are hard to solve.

We discuss hard lattice problems and their application to cryptography in more detail in subsequent sections.

Multivariate Polynomial Cryptography. Another family of problems that is believed to resist quantum computers is related to solving nonlinear equations over a finite field. Cryptosystems that rely on such problems for their security are clubbed under the banner of "multivariate polynomial cryptography" [5–8]. In more detail, the multivariate quadratic polynomial problem, denoted by MQ, is given m quadratic polynomials $f_1, \ldots, f_m$ in n variables $x_1, \ldots, x_n$, with coefficients chosen from a field $\mathbb{F}$, find a solution $\mathbf{z} \in \mathbb{F}^n$ such that $f_i(\mathbf{z}) = 0$ for $i \in [m]$. Evidently, the parameters are chosen so that simple attacks such as linearization do not apply. Indeed, in the worst case, this problem is known to be NP-hard.

The birth of multivariate polynomial cryptography took place in 1988, in an encryption scheme proposed by Matsumoto and Imai [5]. While this scheme was subsequently broken, the general principle found applicability in many subsequent constructions, such as the "Hidden Field Equations" by Patarin [9] or "Unbalanced Oil and Vinegar" [10]. Presently, there exist candidates for secure cryptosystems based on this class of problems that are believed to be quantum secure. We refer the reader to [11] for a detailed survey.

Code-Based Cryptography. Code-based cryptography uses the theory of error-correcting codes to construct cryptosystems. The first candidate of such a cryptosystem was by McEliece [12], based on the hardness of decoding a general linear code, a problem which is known to be NP-hard. To construct the secret key, an error-correcting code is chosen for which an efficient decoding algorithm is known, and which is able to correct up to t errors. The public key is derived from the private key by disguising the selected code as a general linear code. The encryptor generates a codeword using the public key, perturbed by up to t errors. The decryptor recovers the message by performing error correction and efficient decoding of the codeword. The security of the above construction depends heavily on the choice of the error-correcting code used in the construction; to the best of our knowledge, constructions using Goppa codes have remained resilient to attack [13]. Traditionally, the McEliece cryptosystem did not find much deployment due to its large keys and ciphertexts. But there is renewed interest in this family of constructions due to their quantum resilience.

Hash-Based Cryptography. Hash-based cryptography is a general name given to cryptosystems which derive their hardness from hash functions. The simplest and most well-known example of a hash-based cryptosystem is the signature scheme by Merkle [14], which converts a weak signature scheme to a strong one, using hash functions. In more detail, the transformation begins with a signature scheme which is only secure for signing a single message and converts it into a many time signature scheme using the so-called "Merkle tree structure" and by relying only on the existence of hash functions. Since one-time signatures can be based simply on the existence of one-way functions, the security of these constructions is well understood even in the quantum setting. However, the efficiency and generality of hash-based cryptography are restricted, and this limits its popularity.

10.3 Lattice-Based Cryptography

To give the reader a deeper taste of post-quantum cryptography, we focus our attention on lattice-based cryptography for the remainder of this note. To begin, let us define a lattice formally.

Definition 10.1 An *m-dimensional lattice* Λ is a full-rank discrete subgroup of $\mathbb{R}^m$. A *basis* of Λ is a linearly independent set of vectors whose integer linear combinations generate Λ. In cryptography, we are usually concerned with *integer lattices*, i.e., those whose points have coordinates in $\mathbb{Z}^m$.

Among these lattices are the "q-ary" lattices defined as follows: for any integer $q \geq 2$ and any $\mathbf{A} \in \mathbb{Z}_q^{n\times m}$, we define

$$:= \left\{ \mathbf{e} \in \mathbb{Z}^m : \mathbf{A} \cdot \mathbf{e} = \mathbf{0} \bmod q \right\}$$

These lattices are of special interest in cryptography.

The *minimum* distance of a lattice Λ is the length of a shortest nonzero vector:

$$\lambda_1(\Lambda) = \min_{\mathbf{v}\in\Lambda\setminus\{0\}} \|\mathbf{v}\|$$

Here, $\|\cdot\|$ denotes the Euclidean norm. In general, the ith successive minima $\lambda_i(\Lambda)$ is the smallest radius r such that Λ has i linearly independent vectors of norm at most r.

10.3.1 Classic Computational Lattice Problems

In this section, we discuss some classic computational problems over lattices.

Definition 10.2 (*Shortest Vector Problem* (SVP)) Given an arbitrary basis $\mathcal{B}$ of some lattice $\Lambda = \Lambda(\mathcal{B})$, find a nonzero vector $\mathbf{v} \in \Lambda(\mathcal{B})$ such that $\|\mathbf{v}\| = \lambda_1(\Lambda(\mathcal{B}))$.

We note that there is a bound on $\lambda_1(\Lambda(\mathcal{B}))$ by Minkowski's first theorem, which states that for any full-rank lattice $\Lambda(\mathcal{B})$ of rank n,

$$\lambda_1(\Lambda(\mathcal{B})) \leq \sqrt{n}\,(\det(\Lambda(\mathcal{B}))^{\frac{1}{n}}$$

Next, we define the approximate version of this problem. Let $\gamma \geq 1$ be an approximation factor; this is typically taken as a function of the lattice dimension n.

Definition 10.3 (*Approximate Shortest Vector Problem* (SVP_γ)) Given a basis $\mathcal{B}$ of an n-dimensional lattice $\Lambda = \Lambda(\mathcal{B})$, find nonzero vector $\mathbf{v} \in \Lambda(\mathcal{B})$ s.t. $\|\mathbf{v}\| \leq \gamma \cdot \lambda_1(\Lambda(\mathcal{B}))$.

Of particular importance in cryptography is the decision version of the approximate shortest vector problem, which we define next.

Definition 10.4 (*Decisional Shortest SVP* (GapSVP_γ)) Given a basis $\mathcal{B}$ of an n-dimensional lattice and the promise that either $\lambda_1(\Lambda(\mathcal{B})) \leq 1$ or $\lambda_1(\Lambda(\mathcal{B})) \geq \gamma$ determine which is the case.

Definition 10.5 (*Shortest Independent Vector Problem* (SIVP_γ)) Given a basis $\mathcal{B}$ of a full-rank, n-dimensional lattice $\Lambda = \Lambda(\mathcal{B})$, output a set of n linearly independent lattice vectors $S = \{\mathbf{s}_i\}_{i\in[n]}$ s.t. for $i \in [n]$,

$$\|\mathbf{s}_i\| \leq \gamma \cdot \lambda_n(\Lambda(\mathcal{B}))$$

Finally, we define the "bounded distance decoding" problem, which takes as input a lattice Λ and a target point $\mathbf{t}$, with the promise that $\mathbf{t}$ is "close" to Λ, and asks to find the lattice point closest to $\mathbf{t}$.

Definition 10.6 (*Bounded Distance Decoding Problem* (BDD_γ)) Given a basis $\mathcal{B}$ of an n-dimensional lattice $\Lambda = \Lambda(\mathcal{B})$ and a target point $\mathbf{t} \in \mathbb{R}^n$ with the promise that $dist(\Lambda, \mathbf{t}) < d = \lambda_1(\Lambda(\mathcal{B}))/(2\cdot\gamma)$, find the unique lattice point $\mathbf{v}$ such that $\|\mathbf{t} - \mathbf{v}\| < d$.

10.3.1.1 Hardness and Effect on Cryptography

Most of the above problems are known to be NP-hard to solve exactly as well as for sub-polynomial approximation factors. However, cryptographic constructions rely on the hardness of the above problems for polynomial approximation factors, which place them in the realm of NP $\cap$ co-NP. Even for polynomial approximation factors, however, we believe these problems are intractable; indeed, no efficient algorithms are known even for sub-exponential approximation factors despite significant research effort by the community. We refer the reader to [15] for an in-depth discussion.

Early lattice-based cryptosystems such as by Ajtai and Dwork [16], Goldreich, Goldwasser and Halevi [17], and Regev [18] were based on the above problems or variants thereof. While these were important theoretical breakthroughs and introduced ideas that form the cornerstone of lattice-based cryptographic design even today, they were subsequently replaced by simpler systems relying on hardness of a different set of lattice problems, which may be seen as "better suited" for cryptographic design. We discuss these next.

10.3.2 Modern Computational Lattice Problems

Most modern cryptosystems rely on the hardness of the following problems.

10.3.2.1 Short Integer Solution Problem (SIS)

The short integer solution problem was introduced by Ajtai [4] and is defined below.

Definition 10.7 (*Short Integer Solution* ($\mathsf{SIS}_{n,m,q,\beta}$)) Given a uniformly chosen matrix $\mathbf{A} \leftarrow \mathbb{Z}_q^{n\times m}$ and a real-valued parameter β, find a nonzero integer vector $\mathbf{e} \in \mathbb{Z}^m$ s.t.

$$\mathbf{A}\,\mathbf{e} = 0 \mod q \text{ and } \|\mathbf{e}\| \leq \beta$$

Note that the SIS problem can be seen as an average-case short vector problem on the q-ary lattice $\Lambda_q^{\perp}(\text{Å})$ defined above.

Definition 10.8 (*Inhomogeneous Short Integer Solution* ($\mathsf{ISIS}_{n,m,q,\beta}$)) Given a uniformly chosen matrix $\mathbf{A} \leftarrow \mathbb{Z}_q^{n\times m}$, a uniformly chosen vector $\mathbf{u} \leftarrow \mathbb{Z}_q^n$ and a real-valued parameter β, find a nonzero integer vector $\mathbf{e} \in \mathbb{Z}^m$ s.t.

$$\mathbf{A}\,\mathbf{e} = \mathbf{u} \mod q \text{ and } \|\mathbf{e}\| \leq \beta$$

The SIS and ISIS problem can be seen as essentially equivalent, and related to the classic GapSVP problem as follows.

Theorem 10.1 ([4, 19–21]) *For $m = \text{poly}(n)$, any $\beta > 0$, and sufficiently large $q \geq \beta \cdot \text{poly}(n)$, solving the (average case) $\mathsf{SIS}_{n,m,q,\beta}$ (or $\mathsf{ISIS}_{n,m,q,\beta}$) problem with non-negligible probability is at least as hard as solving the decisional approximate shortest vector problem GapSVP_γ and the approximate shortest independent vectors problem SIVP_γ on arbitrary n-dimensional lattices (i.e., in the worst case) with overwhelming probability, for some $\gamma = \beta \cdot \text{poly}(n)$.*

We refer the reader to [15] for a detailed discussion regarding the reductions.

While the SIS and ISIS problem can be used to construct primitives like one-way functions, collision-resistant hash functions and signatures, public-key encryption (and beyond) require the so-called "Learning With Errors" problem LWE [22] or its ring variant RLWE [23]. We define these next.

Definition 10.9 (LWE) Let $q = q(n) \geq 2$ be an integer and let $\chi = \chi(n)$ be a distribution over $\mathbb{Z}$. The $\mathsf{LWE}_{n,q,\chi}$ problem is to distinguish the following two distributions: in the first distribution, sample $(\mathbf{a}_i, b_i)$ uniformly from $\mathbb{Z}_q^{n+1}$. In the second distribution, one first draws $\mathbf{s} \leftarrow \mathbb{Z}_q^n$ uniformly and then samples $(\mathbf{a}_i, b_i) \in \mathbb{Z}_q^{n+1}$ by sampling $\mathbf{a}_i \leftarrow \mathbb{Z}_q^n$ uniformly, $e_i \leftarrow \chi$, and setting $b_i = \langle \mathbf{a}_i, \mathbf{s}\rangle + e_i$. The $\mathsf{LWE}_{n,q,\chi}$ assumption is that the $\mathsf{LWE}_{n,q,\chi}$ problem is infeasible.

We will also need the definition of a B-bounded distribution.

Definition 10.10 (*B-bounded distribution*) A distribution ensemble $(\chi_n)_{n\in\mathbb{N}}$ is called B-bounded if

$$\Pr_{e\leftarrow\chi_n}(\|e\| > B) = \text{negl}(n)$$

Here, $\text{negl}(\cdot)$ refers to a function that decreases faster than the inverse of any polynomial.

Regev [22] proved that for certain moduli q and certain bounded error distributions χ, the $\mathsf{LWE}_{n,q,\chi}$ assumption is true as long as certain worst-case lattice problems are hard to solve using a quantum algorithm. This result was de-quantized by Peikert for exponential modulus [24] and by Brakerski, Langlois, Peikert, Regev, Oded, and Stehlé for polynomial modulus [25].

Theorem 10.2 *For integer dimension n, prime integer q, and integer $B \geq 2n$, there is an efficiently sampleable B-bounded distribution χ such that if there exists an efficient (possibly quantum) algorithm that solves $\mathsf{LWE}_{n,q,\chi}$, then there is an efficient quantum algorithm for solving $\tilde{O}(qn^{1.5}/B)$ approximate worst-case* SIVP *and* GapSVP.

Next, we define the ring variant of the LWE problem, which yields more efficient cryptosystems than LWE.

Definition 10.11 (*Ring Learning With Errors* (RLWE)) Let $f(x) = x^n + 1$ where n is a power of 2. Let $q = q(n)$ be an integer. Let $R = \mathbb{Z}[x]/f(x)$ and let $R_q = R/qR$. Let χ be a probability distribution on R. For $s \in R_q$, let $A_{s,\chi}$ be the probability distribution on $R_q \times R_q$ obtained by choosing an element $a \in R_q$ uniformly at random, choosing $e \leftarrow \chi$ and outputting $(a, a\cdot s + e)$. The decision $\mathsf{RLWE}_{n,q,\chi}$ problem is to distinguish between samples that are either (all) from $A_{s,\chi}$ or (all) uniformly random in $R_q \times R_q$. The $\mathsf{RLWE}_{n,q,\chi}$ assumption is that the $\mathsf{RLWE}_{n,q,\chi}$ problem is infeasible.

Theorem 10.3 ([23]) *Let $r \geq \omega(\sqrt{\log n})$ be a real number and let R, q be as above. Then, there is a randomized reduction from $2^{\omega(\log n)} \cdot (q/r)$ approximate* RSVP *to* $\mathsf{RLWE}_{n,q,\chi}$ *where χ is the discrete Gaussian distribution with parameter r. The reduction runs in time* $\text{poly}(n, q)$.

10.3.2.2 NTRU

Another popular hardness assumption is the **NTRU** assumption defined by [26] which roughly states that it is hard to distinguish a fraction of small elements over R_q from random.

Definition 10.12 ($\mathsf{NTRU}_{q,\chi}$) The NTRU problem $\mathsf{NTRU}_{q,\chi}$ is to distinguish between the following two distributions: in the first distribution sample, a polynomial $h = g/f$ where $f, g \leftarrow \chi$, conditioned on f being invertible in R_q and in the second distribution sample, a polynomial h uniformly over R_q.

Stehlé and Steinfeld [27] showed that the $\mathsf{NTRU}_{q,\chi}$ problem is hard even for unbounded adversaries for χ chosen as the discrete Gaussian distribution with parameter $r > \sqrt{q} \cdot \mathrm{poly}(n)$. However, it is more useful to make the assumption for much smaller $r = \mathrm{poly}(n)$ as in [28].

10.4 Cryptographic Constructions

In this section, we discuss how the aforementioned hardness assumptions can be used to design cryptosystems. Due to space constraints we restrict our attention to the primitive of encryption. We describe the public-key encryption system based on LWE defined by Regev [22].

10.4.1 Public-Key Encryption

Recall the notion of public-key encryption. At a high level, a public-key encryption scheme consists of the following algorithms:

> Setup(1^n): This algorithm takes as input the security parameter (which can be used to fine-tune the efficiency-security trade-off in any construction) and outputs a public key PK and a secret key SK.
> Encrypt(PK, M): This algorithm takes as input public key PK and a message $M \in \{0, 1\}$, and outputs a ciphertext CT.
> Decrypt(PK, SK, CT): This algorithm takes as input the public key PK, the secret key SK, and a ciphertext CT and outputs a message M or $\perp$.

Correctness requires that if (PK, SK) are generated honestly using Setup and CT is generated honestly using Encrypt on inputs (PK, M), then Decrypt(PK, SK, CT) yields M as desired. Security requires that an encryption of M_0 is indistinguishable from an encryption of M_1 for any M_0, M_1.

We proceed to describe a public-key encryption system designed by Regev [22], whose hardness is based on the LWE problem.

Setup(1^n): On input a security parameter n do:

1. Choose a random matrix $\mathbf{A} \in \mathbb{Z}_q^{n \times m}$.
2. Choose a uniformly random $\mathbf{s} \xleftarrow{R} \mathbb{Z}_q^n$.
3. Choose a noise vector $\mathbf{e} \leftarrow \chi^m$.
4. Set $\mathbf{b} = \mathbf{A}^{\mathsf{T}} \cdot \mathbf{s} + \mathbf{e}$.

Output PK $= (\mathbf{A}, \mathbf{b})$ and SK $= \mathbf{s}$.

Encrypt(PK, M): On input public parameters PK and a message $M \in \{0, 1\}$, do:

1. Choose a uniformly random vector $\mathbf{r} \xleftarrow{R} \{0, 1\}^m$.
2. Compute $\mathbf{c}_0 = \mathbf{A} \cdot \mathbf{r}$ and $c_1 = \mathbf{r}^{\mathsf{T}}\mathbf{b} + M\lfloor \frac{q}{2} \rfloor$.

Output the ciphertext CT $:= (\mathbf{c}_0, c_1)$.

Decrypt(PK, SK, CT): On input the public parameters PK, the secret key SK $= \mathbf{s}$ and a ciphertext CT $= (\mathbf{c}_0, c_1)$, do:

1. Let $d = c_1 - \mathbf{c}_0^{\mathsf{T}}\mathbf{s}$.
2. If d is closer to $q/2$ than to 0 output 1, else output 0.

10.4.1.1 Correctness

To see that the encryption scheme is correct, we walk through the steps of decryption:

$$
\begin{aligned}
d &= c_1 - \mathbf{c}_0^{\mathsf{T}}\mathbf{s} \\
&= \left(\mathbf{r}^{\mathsf{T}}\mathbf{b} + M\lfloor \frac{q}{2} \rfloor\right) - \left(\mathbf{A} \cdot \mathbf{r}\right)^{\mathsf{T}}\mathbf{s} \\
&= \mathbf{r}^{\mathsf{T}}(\mathbf{A}^{\mathsf{T}} \cdot \mathbf{s} + \mathbf{e}) + M\lfloor \frac{q}{2} \rfloor - \mathbf{r}^{\mathsf{T}}\mathbf{A}^{\mathsf{T}}\mathbf{s} \\
&= \mathbf{r}^{\mathsf{T}}\mathbf{A}^{\mathsf{T}}\mathbf{s} + \mathbf{r}^{\mathsf{T}}\mathbf{e} + M\lfloor \frac{q}{2} \rfloor - \mathbf{r}^{\mathsf{T}}\mathbf{A}^{\mathsf{T}}\mathbf{s} \\
&= \mathbf{r}^{\mathsf{T}}\mathbf{e} + M\lfloor \frac{q}{2} \rfloor
\end{aligned}
$$

Since $\mathbf{r}$ is binary and $\mathbf{e}$ is chosen from a bounded distribution, it is possible to set the parameters so that $\mathbf{r}^{\mathsf{T}}\mathbf{e}$ is significantly smaller than $q/2$ and can be rounded off to recover the bit M.

10.4.1.2 Security

Security relies on the LWE assumption. Note that by the left-over hash lemma [29], for $m > 2n \log q$ and randomly chosen $\mathbf{r}$, the product $\mathbf{A} \cdot \mathbf{r} = \mathbf{u}$ (say) is uniform. Then, we observe that the ciphertext $(\mathbf{c}_0, c_1)$ is sampled from the LWE distribution as $(\mathbf{u}, \mathbf{u}^{\mathsf{T}}\mathbf{s} + \mathbf{r}^{\mathsf{T}}\mathbf{e} + M\lfloor \frac{q}{2} \rfloor)$, which by the LWE assumption is indistinguishable from uniform $(\mathbf{u}, v)$ which implies that M is hidden.

10.5 Conclusions

We presented a very high-level overview of post-quantum cryptography, with a focus on lattice-based cryptography. This note is too short to contain anything beyond a flavor of the topic of discussion, which is as deep as it is beautiful. We refer the reader to [15] for an excellent survey of lattice-based cryptography and to [3, 13] for more details on post-quantum cryptography at large.

References

1. Goldreich O (2000) Foundations of cryptography: basic tools. Cambridge University Press, New York
2. Shor PW (1994) Algorithms for quantum computation: discrete logarithms and factoring. In: 35th annual symposium on foundations of computer science, Santa Fe, New Mexico, USA, 20–22 Nov 1994, pp 124–134
3. Chen L, Jordan S, Liu YK, Moody D, Peralta R, Perlner R, Smith-Tone D. Report on post-quantum cryptography. https://nvlpubs.nist.gov/nistpubs/ir/2016/nist.ir.8105.pdf
4. Ajtai M (1996) Generating hard instances of lattice problems (extended abstract). In: Proceedings of the twenty-eighth annual ACM symposium on theory of computing, STOC '96
5. Matsumoto T, Imai H (1988) Public quadratic polynomial-tuples for efficient signature-verification and message-encryption. In: Advances in cryptology — EUROCRYPT '88. Springer, Berlin, Heidelberg, pp 419–453
6. Bardet M, Faugere JC, Salvy B, Spaenlehauer PJ (2013) On the complexity of solving quadratic boolean systems. J Complex 29(1):53–75. https://doi.org/10.1016/j.jco.2012.07.001. http://www.sciencedirect.com/science/article/pii/S0885064X12000611
7. Wolf C (2005) Multivariate quadratic polynomials in public key cryptography. PhD thesis, Katholieke Universiteit Leuven
8. Ding J, Yang BY (2009) Multivariate public key cryptography. Springer, Berlin, pp 193–241
9. Patarin J (1996) Hidden fields equations (hfe) and isomorphisms of polynomials (ip): two new families of asymmetric algorithms. In: Maurer U (ed) Advances in cryptology — EUROCRYPT '96
10. Kipnis A, Patarin J, Goubin L (1999) Unbalanced oil and vinegar signature schemes. In: Stern J (ed) Advances in cryptology — EUROCRYPT '99
11. Hashimoto Y (2018) Multivariate public key cryptosystems. In: Mathematical modelling for next-generation cryptography. Springer, pp 17–42
12. McEliece RJ (1978) A Public-key cryptosystem based on algebraic coding theory. Deep space network progress report, vol 44, pp 114–116
13. Overbeck R, Sendrier N (2009) Code-based cryptography
14. Merkle R (1979) Secrecy, authentication and public key systems/a certified digital signature. PhD thesis, Stanford University
15. Peikert C (2016) A decade of lattice cryptography 10:283–424
16. Ajtai M, Dwork C (1997) A public-key cryptosystem with worst-case/average-case equivalence. In: Proceedings of the twenty-ninth annual ACM symposium on theory of computing. ACM, pp 284–293
17. Goldreich O, Goldwasser S, Halevi S (1997) Public-key cryptosystems from lattice reduction problems. In: Annual international cryptology conference. Springer, pp 112–131
18. Regev O (2004) New lattice-based cryptographic constructions. J ACM (JACM) 51(6):899–942
19. Micciancio D, Regev O (2004) Worst-case to average-case reductions based on gaussian measures. SIAM J Comput (SICOMP) 37(1):267–302. Extended abstract in FOCS 2004
20. Gentry C, Peikert C, Vaikuntanathan V (2008) Trapdoors for hard lattices and new cryptographic constructions. In: STOC, pp 197–206
21. Micciancio D, Peikert C (2013) Hardness of sis and lwe with small parameters. In: Crypto
22. Regev O (2009) On lattices, learning with errors, random linear codes, and cryptography. J ACM 56(6). Extended abstract in STOC'05
23. Lyubashevsky V, Peikert C, Regev O (2010) On ideal lattices and learning with errors over rings. In: EUROCRYPT, vol 6110
24. Peikert C (2009) Public-key cryptosystems from the worst-case shortest vector problem. In: STOC, pp 333–342
25. Brakerski Z, Langlois A, Peikert C, Regev O, Stehlé D (2013) Classical hardness of learning with errors. In: Proceedings of the forty-fifth annual ACM symposium on theory of computing, STOC '13. ACM
26. Hoffstein J, Pipher J, Silverman JH (1998) Ntru: a ring-based public key cryptosystem. In: Buhler JP (ed) Algorithmic number theory: third international symposium, ANTS-III Portland, Oregon, USA, 21–25 June 998, Proceedings
27. Stehlé D, Steinfeld R (2011) Making ntru as secure as worst-case problems over ideal lattices. In: Proceedings of the 30th annual international conference on theory and applications of cryptographic techniques: advances in cryptology, EUROCRYPT'11
28. López-Alt A, Tromer E, Vaikuntanathan V (2012) On-the-fly multiparty computation on the cloud via multikey fully homomorphic encryption. In: Proceedings of the forty-fourth annual ACM symposium on theory of computing, STOC '12
29. Barak B, Dodis Y, Krawczyk H, Pereira O, Pietrzak K, Standaert FX, Yu Y (2011) Leftover hash lemma, revisited. In: Annual cryptology conference. Springer, pp 1–20